Breaking Tolerance to Anti-Cancer Cell-Mediated Immunotherapy

IMMUNOTHERAPEUTIC STRATEGIES FOR THE TREATMENT OF GLIOMA

VOLUME 3

Breaking Tolerance to Anti-Cancer Cell-Mediated Immunotherapy Series

Series Editor: Benjamin Bonavida, PhD

Volume 1: Autophagy in Immune Response: Impact on Cancer Immunotherapy
Edited by Salem Chouaib

Volume 2: Immunotherapy in Resistant Cancer: From the Lab Bench Work to Its Clinical Perspectives
Edited by Jorge Morales-Montor and Mariana Segovia-Mendoza

Volume 3: Immunotherapeutic Strategies for the Treatment of Glioma
Edited by Christopher Jackson and Michael Lim

Upcoming Volumes:

Breaking Tolerance to Unresponsiveness to Immunotherapy by Natural Killer Cells
Edited by Sandro Matosevic

NK Cells in Cancer Immunotherapy: Successes and Challenges
Edited by Anahid Jewett

Breaking Tolerance to Pancreatic Cancer Unresponsive to Immunotherapy
Edited by Kasuya Hideki and Itzel Bustos Villalobos

Breaking Tolerance to Check Point Inhibitors
Edited by Prashanth Prithviraj

Breaking Tolerance to Anti-Cancer Cell-Mediated Immunotherapy

IMMUNOTHERAPEUTIC STRATEGIES FOR THE TREATMENT OF GLIOMA

VOLUME 3

Edited by

CHRISTOPHER JACKSON, M.D.

Assistant Professor of Neurosurgery at Johns Hopkins University School of Medicine, Johns Hopkins Hospital, Baltimore, MD, United States

MICHAEL LIM, M.D.

Professor and Chair, Department of Neurosurgery; Professor, by courtesy, of Medicine (Oncology), of Neurology, of Otolaryngology, and of Radiation Oncology, Stanford University School of Medicine, Center for Academic Medicine, Palo Alto, CA, United States

ELSEVIER

ACADEMIC PRESS

An imprint of Elsevier

Academic Press is an imprint of Elsevier
125 London Wall, London EC2Y 5AS, United Kingdom
525 B Street, Suite 1650, San Diego, CA 92101, United States
50 Hampshire Street, 5th Floor, Cambridge, MA 02139, United States
The Boulevard, Langford Lane, Kidlington, Oxford OX5 1GB, United Kingdom

Notices
Knowledge and best practice in this field are constantly changing. As new research and experience broaden our understanding, changes in research methods, professional practices, or medical treatment may become necessary.

Practitioners and researchers must always rely on their own experience and knowledge in evaluating and using any information, methods, compounds, or experiments described herein. In using such information or methods they should be mindful of their own safety and the safety of others, including parties for whom they have a professional responsibility.

To the fullest extent of the law, neither the Publisher nor the authors, contributors, or editors, assume any liability for any injury and/or damage to persons or property as a matter of products liability, negligence or otherwise, or from any use or operation of any methods, products, instructions, or ideas contained in the material herein.

Library of Congress Cataloging-in-Publication Data
A catalog record for this book is available from the Library of Congress

British Library Cataloguing-in-Publication Data
A catalogue record for this book is available from the British Library

ISBN: 978-0-12-819755-4

For information on all Academic Press publications
visit our website at https://www.elsevier.com/books-and-journals

Publisher: Stacy Masucci
Senior Acquisitions Editor: Rafael Teixeira
Editorial Project Manager: Samantha Allard
Production Project Manager: Punithavathy Govindaradjane
Cover Designer: Victoria Pearson

Typeset by STRAIVE, India

Cover Image Insert

Multiplex immunofluorescence of glioblastoma tissue highlights the complexity of the immune microenvironment. *Magenta = tumor cells, Cyan = PD-1, Orange = macrophages (CD68), Red = Tregs (FoxP3), Yellow = cytotoxic T cells (CD8), Green = PD-L1, Blue = DAPI.*

Aims and scope of series "Breaking Tolerance to Anti-Cancer Cell-Mediated Immunotherapy"

The role of the immune system in the eradication of cancers has been investigated for several decades with controversial findings. The controversy was the result of a poor understanding of the underlying mechanisms that govern responsiveness and unresponsiveness. Hence, significant advances have been made with respect to the regulation of the host immune response against cancer, and several immunotherapeutics have been recently introduced and used clinically. These include both antibody- and cell-mediated immunity targeting the cancer cells. Such immunotherapies led to significant clinical responses in various cancer types that were unresponsive to conventional therapies. However, only a subset of cancer patients respond to such immunotherapeutics, and also there is a responding subset that develops resistance to further treatment. Various studies have examined the potential underlying mechanisms involved in resistance and identified a variety of gene products that play pivotal roles in maintaining the resistant phenotype of the cancer cells to cell-mediated immunotherapy.

The main objective of the proposed series "Breaking Tolerance to Anti-Cancer Cell-Mediated Immunotherapy" is the development of individual volumes that are focused on the application of particular sensitizing agents that, when used in combination with cell-mediated immunotherapy, result in the reversal of resistance.

A variety of different classes of immunosensitizing agents has been reported. Each individual volume will focus on one class of immunosensitizing agents and their effects on the reversal of cell-mediated immune resistance in different cancers. Emphasis will be on biochemical, molecular, and genetic mechanisms by which the sensitizing agents mediate their effects individually and/or in combination with immunotherapy. Each editor will compile nonoverlapping review chapters on the therapeutic role of specific sensitizing agents used in combination with conventional immunotherapy and the reversal of resistance. There will also be an emphasis on discrimination of responses obtained in various cancer types.

The scope of the series is to provide updated information to scientists and clinicians that is valuable in their quest to gather information, carry out new investigations, and develop novel immunosensitizing agents that are both more potent and also might be active whereby the existing ones were not active.

Dr. Benjamin Bonavida, PhD (Series Editor)

About the Series Editor

Dr. Benjamin Bonavida, PhD, (Series Editor), is currently Distinguished Research Professor at the University of California, Los Angeles (UCLA). He is affiliated with the Department of Microbiology, Immunology and Molecular Genetics, UCLA David Geffen School of Medicine. His research career, thus far, has focused on investigations in the fields of basic immunochemistry and cancer immunobiology. His research investigations have ranged from the biochemical, molecular, and genetic mechanisms of cell-mediated killing and tumor cell resistance to chemo-immuno cytotoxic drugs. The reversal of tumor cell resistance was investigated by the use of various selected sensitizing agents based on molecular mechanisms of resistance. In these investigations, there was the newly characterized dysregulated NF-κB/Snail/YY1/RKIP/PTEN loop in many cancers that was reported to regulate cell survival, proliferation, invasion, metastasis, and resistance. Emphasis was focused on the roles of the tumor suppressor Raf Kinase Inhibitor Protein (RKIP), the tumor promoter Yin Yang 1 (YY1), and the role of nitric oxide as a chemo-immuno-sensitizing factor. Many of the earlier mentioned studies are centered on the clinical challenging features of cancer patients' failure to respond to both conventional and targeted therapies.

The Editor has been active in the organization of regular sequential international miniconferences that are highly focused on the roles of YY1, RKIP, and nitric oxide in cancer and their potential therapeutic applications. Several books edited or coedited by the Editor have been published. In addition, the Editor has been the Series Editor of books (over 23) published by Springer on *Resistance to Anti-Cancer Targeted Therapeutics*. In addition, the Editor is presently the Series Editor of three series published by Elsevier/Academic Press on *Cancer Sensitizing Agents for Chemotherapy*, *Breaking Tolerance to Anti-Cancer Cell-Mediated Immunotherapy*, and *Breaking Tolerance to Anti-Cancer Immunotherapy*. Lastly, the Editor is the Editor-in-Chief of the Journal Critical Reviews in Oncogenesis. The Editor has published over 500 research publications and reviews in various scientific journals of high impact.

Acknowledgments: The Series Editor acknowledges the Department of Microbiology, Immunology, and Molecular Genetics and the UCLA David Geffen School of Medicine for their continuous support. The Series Editor also acknowledges the assistance of Mr. Rafael Teixeira, Acquisitions Editor, for Elsevier/Academic Press, and the excellent assistance and support of Ms. Samantha Allard, Editorial Project Manager, for Elsevier/Academic Press, for their continuous cooperation throughout the development of this book.

Aims and Scope of Volume

The aim of this volume is to highlight recent advances and state-of-the-art strategies driving the development of immunotherapy for brain tumors. The clinical implementation of immunotherapy for cancer—born out of fundamental discoveries in how solid tumors interact with the immune system—is one of the most impactful breakthroughs in the history of oncology. Many patients with advanced cancers are now surviving years-beyond what would be expected with traditional therapies. Despite these encouraging results, a minority of patients with immunotherapy-sensitive pathologies derive a survival benefit, and some pathologies only rarely and sporadically respond. The field's focus is now on optimizing the implementation of available therapeutics as well identifying novel immunosuppressive mechanisms that will give rise to the next generation of immunotherapeutic targets. In some instances, combination immunotherapy can boost response rates to immunotherapy-sensitive tumors. For example, concurrent administration of cytotoxic T lymphocyte-4 (CTLA-4)- and programmed death-1 (PD-1)-blocking antibodies is more effective than either monotherapy in subtypes of unresectable NSCLC, mesothelioma, metastatic melanoma, renal cell carcinoma, hepatocellular carcinoma, and colorectal cancer with microsatellite instability. As new immunotherapies, including antibodies targeting alternative immune checkpoints, adoptive cell therapy, and other approaches, are added to the armamentarium, there is reason for optimism that response rates in these pathologies will continue to improve.

Some immunotherapy-unresponsive pathologies, however, may require more nuanced approaches. Malignant gliomas present many unique challenges to immunotherapy based on their location within the CNS, a dearth of high-quality mutations and targetable neoantigens, low immune cell density and profound exhaustion of infiltrating lymphocytes, systemic immunosuppression due to T-cell sequestration and tolerance, and iatrogenic immunosuppression associated with standard-of-care chemotherapy and radiation. Our book focuses on how emerging immunotherapeutic strategies address these unique aspects of glioma immunobiology. This book is by no means an exhaustive list of promising or emerging immunotherapies for glioma, but rather a compilation of topics we believe to be illustrative of the most significant challenges currently facing the field. The scope of this book is to provide current information to scientists and clinicians that will inform and focus their efforts to make immunotherapy for glioma a reality.

About the Volume Editors

Christopher Jackson, M.D., graduated Summa Cum Laude from Wake Forest University with a Bachelor of Arts in English Literature and minors in Chemistry and Neuroscience in 2008. During his undergraduate studies, Dr. Jackson worked in the laboratory of Dr. Susan Fahrbach examining the physiology of the honeybee mushroom bodies during learning and memory tasks. This work was recognized with a Barry M. Goldwater Scholarship in 2007. He matriculated to the Johns Hopkins University School of Medicine in 2008 and began working in the tumor immunology laboratory with Drs. Michael Lim, Charles Drake, and Drew Pardoll studying mechanisms by which brain tumors suppress local and systemic immune responses as well as developing combination immunotherapy strategies to overcome brain tumor-mediated immune tolerance. During this time, he completed a Howard Hughes research fellowship in the laboratories of Dr. Charles Drake and Michael Lim. As a medical student, he was recognized with the Harry C. Saltzstein Prize for Medical Writing as well as the Harvey Cushing Medical Student Research Award. Dr. Jackson completed residency in Neurosurgery at Johns Hopkins Hospital in 2020 and joined the faculty as an Assistant Professor. He has authored over 40 peer-reviewed publications to date and runs an active laboratory focused on brain tumor immunotherapy and the role of inflammation in cerebrovascular disease.

Dr. Jackson has extensive experience in the field of cancer immunology and is an expert on brain tumor-mediated immunosuppression. He is an Assistant Professor in the Department of Neurosurgery at the Johns Hopkins University School of Medicine.

Dr. Michael Lim is a Professor and Chair of Neurosurgery at Stanford University. Dr. Lim obtained his MD from the Johns Hopkins University School of Medicine. He then completed his residency in Neurosurgery at Stanford University Hospital and went on to become a Professor of Neurosurgery, Oncology, and Radiation Oncology at Johns Hopkins before returning to Stanford as Chair. Dr. Lim's surgical interest is in both benign and malignant brain tumors, with a particular interest in gliomas (including ependymoma), meningioma, pituitary tumors, and skull base tumors. He has extensive experience in new and innovative neurosurgical techniques, including image-guided surgery, microsurgery, minimally invasive procedures, and endoscopic surgery.

Dr. Lim's primary research interest is developing immune-based therapies against brain tumors. His research laboratory is focused on understanding the mechanisms of immune evasion by primary brain tumors. Findings from his laboratory are directed toward translation to novel therapies against brain tumors. In addition to running a laboratory, he currently serves as the principal investigator of several large brain tumor immunotherapy clinical trials based on findings from his laboratory.

Preface

The quest to cure brain tumors with immunotherapy is riddled with challenges, many of which are unique consequences of glioma biology, the immune environment of the central nervous system, and current treatment paradigms. Accordingly, the successes of immunotherapy that have been achieved in many other malignancies have yet to be realized for patients with brain tumors. This book addresses select topics representing some of the most significant impediments to brain tumor immunotherapy identified to date and is intended to serve as a framework for understanding state-of-the-art strategies aimed at overcoming these obstacles. This book will be a valuable resource for researchers and clinicians in the fields of immunology, immuno-oncology, and neuro-oncology seeking to understand the status and direction of the field, the challenges we face, and many of the pivotal discoveries that brought us to this point.

Immunosuppression in patients with high-grade gliomas is mediated by multiple mechanisms and affects both local and systemic immune functions. The tumor microenvironment is replete with immunosuppressive cytokines, has a high density of suppressive myeloid cells, and stimulates upregulation of immune checkpoints. In addition, the blood-brain barrier tightly controls ingress of therapeutics and immune cells into the brain, particularly along the invading edges of the tumor. Antigen processing and presentation is unique in the brain and skews away from cytotoxic immune responses. Furthermore, brain tumors promote local and systemic tolerance against tumor antigens, and lymphocytes infiltrating brain tumors are characterized by a severely exhausted phenotype. Finally, the most effective chemotherapy and radiation regimens against high-grade gliomas are profoundly immunosuppressive.

The chapter "Overcoming immune suppression in glioblastoma" delineates the factors that contribute to immunosuppression in brain tumor patients and discusses strategies for restoring immune function. Impaired immune function has been identified as a negative prognostic factor for patients with glioblastoma undergoing standard chemotherapy and radiation. Developing protocols that prevent immunosuppression and restore immune function after chemotherapy and radiation is a critical step toward realizing the potential of immunotherapy. In this way, the chapter by Drs. Grossman and Ellsworth sets the stage for the subsequent chapters on specific immunotherapeutic strategies.

If preventing immunosuppression is analogous to leveling the battlefield, then immunogenomics is identifying the enemy. One of the major challenges in generating an immune response against high-grade gliomas is that the low mutational burden of these tumors corresponds with few high-quality neoantigens that distinguish neoplastic from normal tissue. Immunogenomics leverages high-throughput sequencing technologies to identify and characterize tumor-specific neoantigens and quantify their immunogenicity. The chapter "Applied cancer immunogenomics in glioblastoma" by Dr. Dunn and colleagues discusses immunogenomics as a critical tool for elucidating the nuances of interactions between brain tumor antigens and immune cells. These data will be instrumental in increasing the potency and specificity of the next generation of immunotherapies against glioblastoma. In parallel, platforms must be developed that prime these immune responses, effectively rendering "cold" tumors "hot." Dr. Bloch and colleagues explore the utility of one such approach that has demonstrated promise by using heat shock protein vaccines to prime the tumor microenvironment and facilitate antigen cross-presentation.

The paucity of antitumor immune cells in the glioblastoma microenvironment presents a major impediment to the effectiveness of immunotherapies designed to invigorate immune cells against tumors. Adoptive cell therapy involves infusing tumor-targeting immune cells into patients in order to increase the number of cells available to mediate antitumor immune responses. In the chapter "Adoptive cell therapy for glioma", Drs. Hoang-Minh and Mitchell discuss the rationale for adoptive cell therapy in this context as well as specific protocols for harvesting, activating, and infusing antitumor immune cells so as to optimize tumor antigen-specific responses. They conclude that the potential of adoptive cell therapy to increase antitumor immune cell density within gliomas makes this approach particularly promising for combination with other therapies that activate immune cells or prevent exhaustion.

Immune checkpoints are expressed on activated immune cells and function as a "brake" to mitigate overly robust inflammatory responses and prevent autoimmunity. Many solid tumors, including high-grade gliomas, constitutively express immune checkpoint ligands, usurping these pathways to obviate immune attack. The clinical development of monoclonal antibodies that block interactions between immune checkpoint molecules and their ligands has been the most significant breakthrough in cancer immunotherapy; however, a series of phase III trials of immune checkpoint blockade in glioblastoma failed to meet their primary endpoints. In the chapter "Immune checkpoint blockade therapy in high-grade glioma," Dr. Fecci and colleagues discuss the nuances of these trials and how these findings are being used to inform the next iteration of immune checkpoint inhibitor-based strategies against glioblastoma. They also discuss specific impediments to the effectiveness of immune checkpoint blockade in in this context and strategies for overcoming these obstacles. The authors conclude that, despite the early disappointments with immune checkpoint blockade monotherapy, these agents are likely to play a critical role in an effective combination immunotherapy regimen as the key mechanism to fight T-cell exhaustion. Other approaches will be required, however, to address the availability and activation status of tumor antigen-directed T cells.

Although high-grade gliomas have a very low density of T cells, they exhibit a high density of myeloid cells. In fact, myeloid cells are the second most frequent cell type in glioblastomas after tumor cells. The majority of these cells function as myeloid-derived suppressor cells, which promote tumor growth by impeding antitumor immune responses. Myeloid cells also function as the primary antigen-presenting cells and serve as a hub for initiating adaptive immune responses. In this context, dendritic cell function is a focus of several immunotherapeutic efforts, including infusion of dendritic cells pulsed with tumor antigens and administration of TLR agonists. Dr. Okada and colleagues discuss the multitude and variety of roles played by myeloid cells in initiating, sustaining, and dampening immune responses against brain tumors. The authors' discussion of myeloid cells in glioblastoma strongly supports the argument that a breakthrough in immunotherapy against glioblastoma will require agents that effectively target myeloid cell populations.

The discovery that the indoleamine 2,3-dioxygenase 1 (IDO), an enzyme involved in tryptophan metabolism, is expressed by tumor-infiltrating immune cells and correlates with a significant decrease in glioblastoma patient survival is one of the most well-developed examples of the intersection between immunology and metabolism. In the chapter "Indoleamine 2,3-dioxygenase 1 (IDO): a mediator of immunoresistance in adults with brain cancer treated with immunomodulatory therapy," Dr. Wainwright and colleagues detail the evidence for IDO as a novel immunosuppressive pathway and discuss the pleotropic effects of this enzyme in health and disease. They conclude with an insightful view into the future of IDO research and the potential it holds, which is a shining example within the burgeoning field of immunometabolism.

Integrating immunotherapeutic approaches with emerging antitumor technologies presents a significant opportunity to drive the field forward. This must be done in real time as each new therapy emerges rather than retroactively after a new agent has been optimized as a monotherapy (but potentially not for optimizing synergy with immunotherapy). Drs. Li and Wu discuss the potential of targeted therapies against chemokine receptors in combination with existing immunotherapies such as immune checkpoint inhibitors. Such an approach may also allow for the development of combination regimens that attack multiple aspects of tumorigenesis and progression without compromising the host immune response.

The compilation of expertise in this book serves as an up-to-date window into the challenges, strategies, and ultimately promise of immunotherapy for brain tumors. It is our hope that this book serves as a platform that informs and inspires the next breakthrough.

Christopher Jackson
Michael Lim

Contents

6. Immune checkpoint blockade therapy in high-grade glioma

Jessica Waibl Polania, Selena Lorrey, Daniel Wilkinson, and Peter E. Fecci

7. Immunomodulatory roles of myeloid cells in gliomas

Pavlina Chuntova, Payal B. Watchmaker, Ryan Gilbert, and Hideho Okada

8. Indoleamine 2,3-dioxygenase 1 (IDO): A mediator of immunoresistance in adults with brain cancer treated with immunomodulatory therapy

Lijie Zhai, Erik Ladomersky, Kristen L. Lauing, Lakshmi Bollu, April Bell, Miri Kim, Erik Rabin, and Derek A. Wainwright

9. Chemokine receptor antagonists as immunotherapy agents and adjuncts for glioblastoma

Adela Wu, Michael Jin, and Gordon Li

Contributors

April Bell Department of Neurological Surgery, Northwestern University Feinberg School of Medicine, Chicago, IL, United States

Orin Bloch Department of Neurological Surgery, University of California, Davis, Sacramento, CA, United States

Lakshmi Bollu Department of Neurological Surgery, Northwestern University Feinberg School of Medicine, Chicago, IL, United States

E. Antonio Chiocca Department of Neurosurgery, Brigham and Women's Hospital, Harvard Medical School, Boston, MA, United States

Pavlina Chuntova Department of Neurological Surgery, University of California, San Francisco, CA, United States

Andrew T. Coxon Andrew M. and Jane M. Bursky Center for Human Immunology and Immunotherapy Programs, Washington University School of Medicine, St. Louis, MO, United States

Gavin P. Dunn Department of Neurological Surgery; Andrew M. and Jane M. Bursky Center for Human Immunology and Immunotherapy Programs, Washington University School of Medicine, St. Louis, MO, United States

Susannah G. Ellsworth University of Pittsburgh Hillman Cancer Center, Pittsburgh, PA, United States

Peter E. Fecci Duke Brain Tumor Immunotherapy Program, Department of Neurosurgery, Duke University Medical Center; Department of Pathology, Duke University Medical Center; Department of Immunology, Duke University Graduate School, Durham, NC, United States

Ryan Gilbert Department of Neurological Surgery, University of California, San Francisco, CA, United States

Stuart A. Grossman Johns Hopkins University School of Medicine, Baltimore, MD, United States

Aden P. Haskell-Mendoza Department of Neurological Surgery, University of California, Davis, Sacramento, CA, United States

Lan B. Hoang-Minh Lillian S. Wells Department of Neurosurgery, UF Brain Tumor Immunotherapy Program, Preston A. Wells, Jr. Center for Brain Tumor Therapy, McKnight Brain Institute, University of Florida, Gainesville, FL, United States

Michael Jin Department of Neurosurgery, Stanford University School of Medicine, Stanford, CA, United States

Miri Kim Department of Neurological Surgery, Loyola University Chicago Medical Center, Maywood, IL, United States

Erik Ladomersky Department of Neurological Surgery, Northwestern University Feinberg School of Medicine, Chicago, IL, United States

Kristen L. Lauing Department of Neurological Surgery, Northwestern University Feinberg School of Medicine, Chicago, IL, United States

Gordon Li Department of Neurosurgery, Stanford University School of Medicine, Stanford, CA, United States

Connor J. Liu Department of Neurological Surgery; Andrew M. and Jane M. Bursky Center for Human Immunology and Immunotherapy Programs, Washington University School of Medicine, St. Louis, MO, United States

Selena Lorrey Duke Brain Tumor Immunotherapy Program, Department of Neurosurgery, Duke University Medical Center; Department of Immunology, Duke University Graduate School, Durham, NC, United States

Duane A. Mitchell Lillian S. Wells Department of Neurosurgery, UF Brain Tumor Immunotherapy Program, Preston A. Wells, Jr. Center for Brain Tumor Therapy, McKnight Brain Institute, University of Florida, Gainesville, FL, United States

Hideho Okada Department of Neurological Surgery, University of California; The Parker Institute for Cancer Immunotherapy; Cancer Immunotherapy Program, University of California, San Francisco, CA, United States

Jessica Waibl Polania Duke Brain Tumor Immunotherapy Program, Department of Neurosurgery; Department of Pathology, Duke University Medical Center, Durham, NC, United States

Erik Rabin Department of Neurological Surgery, Northwestern University Feinberg School of Medicine, Chicago, IL, United States

Isaac H. Solomon Department of Pathology, Brigham and Women's Hospital, Harvard Medical School, Boston, MA, United States

Derek A. Wainwright Department of Neurological Surgery, Northwestern University Feinberg School of Medicine; Department of Medicine, Division of Hematology and Oncology, Northwestern University Feinberg School of Medicine; Robert H. Lurie Comprehensive Cancer Center, Northwestern University; Department of Microbiology-Immunology, Northwestern University Feinberg School of Medicine, Chicago, IL, United States

Payal B. Watchmaker Department of Neurological Surgery, University of California, San Francisco, CA, United States

Daniel Wilkinson Duke Brain Tumor Immunotherapy Program, Department of Neurosurgery, Duke University Medical Center, Durham, NC, United States

Adela Wu Department of Neurosurgery, Stanford University School of Medicine, Stanford, CA, United States

Lijie Zhai Department of Neurological Surgery, Northwestern University Feinberg School of Medicine, Chicago, IL, United States

CHAPTER

1

Mechanisms of immune suppression in glioblastoma

Susannah G. Ellsworth[a] *and Stuart A. Grossman*[b]

[a]University of Pittsburgh Hillman Cancer Center, Pittsburgh, PA, United States [b]Johns Hopkins University School of Medicine, Baltimore, MD, United States

Abstract

Despite advances in immunotherapy for other solid tumors, glioblastoma multiforme (GBM), the most common malignant primary brain tumor, has thus far demonstrated resistance to novel immunotherapeutics such as immune checkpoint inhibitors. In GBM, resistance to immunotherapy may be partly attributable to the immunosuppression that is characteristic of this disease. Both tumor-driven and iatrogenic mechanisms of immune suppression affect systemic immune status as well as the tumor microenvironment in patients with GBM. A thorough understanding of the pathophysiology of immunosuppression in GBM is needed to develop strategies to overcome resistance to immunotherapy in this disease. This chapter summarizes GBM-induced changes in circulating immune cell populations and in the tumor microenvironment and discusses how commonly used interventions in GBM patients, including radiation, chemotherapy, and corticosteroids, also result in immunosuppression.

Abbreviations

ANG-2	angiopoietin 2
BBB	blood-brain barrier
CCL-2	chemokine ligand 2
CD	cluster of differentiation
CNS	central nervous system
CSF	cerebrospinal fluid
CTLA	cytotoxic-T-lymphocyte-associated antigen
CXCL	chemokine C-X-C ligand
EGF(R)	epidermal growth factor (receptor)
GBM	glioblastoma
Gy	Gray
HIF1-α	hypoxia-inducible factor 1α
HLA	human leukocyte antigen
ICI	immune checkpoint inhibitor
IDH	isocitrate dehydrogenase
IDO	indoleamine 2,3-deooxygenase

Immunotherapeutic Strategies for the Treatment of Glioma
https://doi.org/10.1016/B978-0-12-819755-4.00016-3

IFN	interferon
IL	interleukin
IV	intravenous
LAG-3	lymphocyte activation gene 3
MCP-1	monocyte chemoattractant protein 1
MDSC	myeloid-derived suppressor cells
MGMT	methylguanine methyltransferase
MHC	major histocompatibility complex
MRI	magnetic resonance imaging
NF-κB	nuclear factor kappa-light chain enhancer of activated B cells
NK	natural killer
PD-(L)1	programmed death (ligand) 1
PHA	phytohemagglutinin
STAT3	signal transduction and activator of transcription 3
TAM	tumor-associated macrophage
TGF	transforming growth factor
TIGIT	T cell immunoreceptor with Ig and ITIM domains
TIL	tumor-infiltrating lymphocyte
TIM	tumor-infiltrating myeloid cell
TME	tumor microenvironment
Treg	regulatory T cell
VEGF	vascular endothelial growth factor

Conflict of interest

No potential conflicts of interest were disclosed.

Introduction

Despite initial optimism about the potential of immune checkpoint blockade to improve dismal outcomes in patients with glioblastoma (GBM), three recent phase III trials that enrolled over 1000 patients failed to demonstrate a benefit to checkpoint inhibitor-based immunotherapy in GBM. Checkmate-143, which tested nivolumab vs bevacizumab in 369 patients with recurrent disease, a high proportion of whom were steroid-dependent, demonstrated no difference in overall survival, lower response rates (7.8% vs 23.1%), and a shorter progression-free survival time in patients who received nivolumab, although a suggestion of durable response was present in the small proportion (8%) of patients who did obtain a response to nivolumab [1]. In the phase III study Checkmate-498, 500 patients were randomized to receive radiotherapy with either concurrent temozolomide (standard of care) or concurrent nivolumab [2]. Patients were selected based on a lower likelihood of response to temozolomide based on MGMT-unmethylated status. Likewise, this study failed to meet its primary endpoint (overall survival) or to demonstrate improvements in the secondary endpoints of progression-free survival or 2-year overall survival. Finally, Checkmate-548 randomized patients to standard concurrent temozolomide-based chemoradiation with or without the addition of nivolumab [3]. While final results of the study have not been released as of this writing, preliminary analysis showed that the study's primary endpoint of improved progression-free survival was not met. The disappointing outcomes of these studies demonstrate that much more work needs to be done to successfully harness an immune response against GBM. A thorough understanding of the manifold ways in which GBM depresses antitumor immunity is necessary in order to design strategies to overcome the multifactorial immunosuppression induced by this disease. The following chapter will review both intrinsic and treatment-related mechanisms of immunosuppression in GBM (Fig. 1).

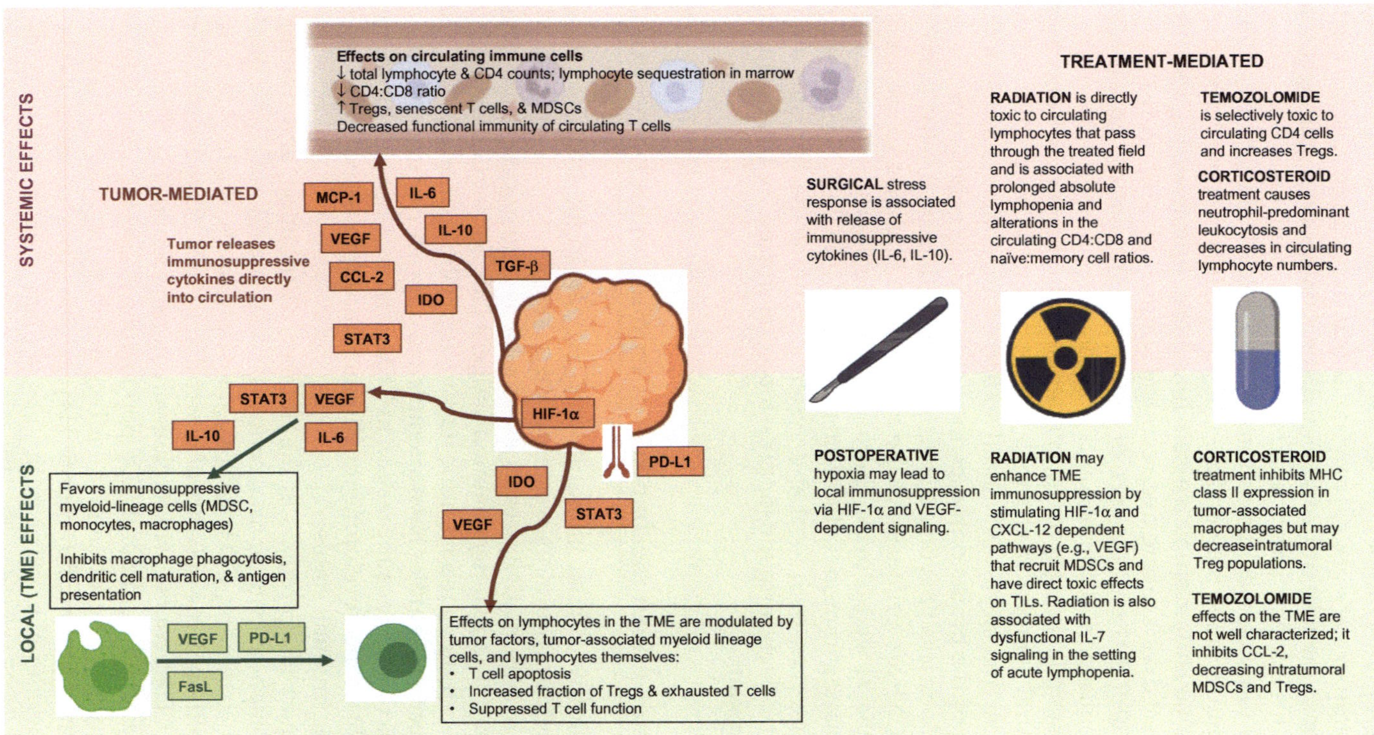

FIG. 1 Mechanisms of immune suppression in glioblastoma.

Past dogma held that the central nervous system (CNS) was an "immune-privileged" site due to a lack of draining lymphatics and the presence of the blood-brain barrier. In reality, the CNS has a complex network of lymphatics that drain directly to cervical lymph nodes, and circulating immune cells utilize several distinct pathways to gain access to the CNS and augment the brain's intrinsic immune surveillance mechanisms. While drainage of cerebrospinal fluid (CSF) into cervical lymph nodes was described as early as 1993 [4], no clear pathway for communication between the CSF and peripheral lymphatic channels was identified until 2015, when lymphatic vessels lining the dura itself [5] and the dural venous sinuses, with direct drainage into deep cervical lymph nodes [6], were described in rodents. It was not until 2019 that the presence of intradural lymphatic channels was confirmed in humans and nonhuman primates and visualized noninvasively using MRI [7]. The "glymphatic" system is thought to drain interstitial fluid from the brain parenchyma itself into the CSF; dural lymphatic channels then provide a direct connection between the glymphatic system and extra-CNS nodal stations. Our understanding of the complex immune landscape of the brain continues to evolve, and an exhaustive summary of normal CNS immune responses is beyond the scope of this chapter. Briefly, surveillance immune cells within the CNS are primarily of the macrophage lineage and are found in the parenchyma as microglia, as well as in the choroid plexus, meninges, and surrounding CNS blood vessels [8]. Circulating immune cells are also recruited into the CNS in instances of infection or trauma [9]. However, both local and systemic immune mechanisms are suppressed by GBM, and tumor-driven immunosuppression is reinforced by treatment-induced toxicity to the immune system.

Tumor-driven mechanisms of immunosuppression in GBM

Reports of defective cellular immunity in GBM date to the 1970s, when serum from patients with primary brain tumors was reported to suppress T-cell-modulated processes, including delayed hypersensitivity reactions, as well as lymphocyte proliferation in response to phytohemagglutinin and other common antigens [10]. Subsequent reports confirmed multiple deficits in T-cell-mediated immune responses in GBM patients [11]. To date, thousands of clinical and preclinical studies have attempted to improve our understanding of the underlying mechanisms of immunosuppression in GBM, which co-opts multiple immune signaling pathways to facilitate the development of an immunosuppressive tumor microenvironment and alter systemic immune status.

GBM causes marked alterations in the number, distribution, and function of circulating immune cells. Compared with normal controls, patients with malignant glioma exhibit relative decreases in circulating CD3+ and CD4+ populations, leading to a decreased CD4+:CD8+ ratio [12], although the majority of newly diagnosed GBM patients maintain peripheral lymphocyte and CD4+ counts within clinically normal limits [13]. The proportion of Tregs within circulating and intratumoral CD4 populations is also higher in glioma patients compared with healthy controls, although total Treg numbers may be decreased in some individuals due to overall lower CD4 counts [14]. Higher Treg concentrations correlate with decreased CD4 proliferation capacity in vitro, a phenomenon that reverses when cultures are depleted of CD25+ Treg cells [15]. Compared with CD25-negative T cells, Tregs demonstrate high expression of CTLA-4, as well as an increased production of the immunosuppressive cytokine IL-10 and a decreased production of immunostimulatory cytokines such as IL-2 and IFN-γ following PMA stimulation [16]. Enrichment of lymphocyte populations with Tregs may therefore explain prior observations of decreased production of IL-2 and decreased IL-2 receptor expression, as well as diminished intracellular signaling in response to T-cell receptor stimulation, among lymphocytes isolated from GBM patients [17, 18]. An increased proportion of senescent T cells has also been identified in peripheral lymphocytes obtained perioperatively in GBM patients [19]. The loss of circulating T cells has been partially attributed to marrow sequestration due to the loss of sphingosine 1-phosphate-1 (S1P1) function [20]. In addition to abnormalities among circulating lymphoid-lineage cells, higher levels of circulating (CD33+/HLADR-) myeloid-derived suppressor cells (MDSC) have been reported in GBM patients compared with age-matched healthy controls [21] and with patients undergoing intracranial surgery for benign indications [22]. T-cell dysfunction in GBM may be partly attributable to increased MDSC populations, which are thought to suppress T-cell activity through metabolic alterations, including arginine depletion [23].

Molecular genomic techniques have permitted highly detailed characterization of the tumor immune microenvironment in GBM, whose primary cellular components include tumor-infiltrating lymphocytes [CD4, CD8, and regulatory T (Treg) cells, as well as dendritic and natural killer (NK) cells], tumor-associated macrophages and microglia (TAMs), and MDSCs. In an analysis of the Cancer Genome Atlas, Thorsson et al. found that gliomas tended to express either "lymphocyte depleted" (C4) or "quiet" (C5) immune phenotypes, with increased macrophage-to-lymphocyte ratios compared with other categories [24]. While C4 tumors had a significantly worse prognosis compared with other groups, C5 tumors did not (despite having the highest macrophage-to-lymphocyte ratio of all tumors, typically thought to be a negative prognostic factor). This may be due to enrichment of the C5 group with low-grade gliomas, particularly isocitrate dehydrogenase (IDH)-mutant tumors, which have a favorable prognosis and have previously been demonstrated to have reduced populations of macrophages, microglia, monocytes, and polymorphonuclear leukocytes compared with IDH-wild-type tumors [25]. Mass cytometry has also provided an insight into changes in both circulating and tumor-infiltrating immune phenotypes over the course of illness in GBM; for example, Alban et al. reported that prolonged survival in GBM patients was associated with decreased circulating MDSC populations and that MDSC infiltration of tumor at the time of recurrence was also a negative prognostic biomarker [22].

Emerging data further highlight the role of tumor-infiltrating myeloid cells in shaping an immunosuppressive tumor microenvironment. Recently, it was reported that the majority of PD-L1 expression in the TME was not in tumor cells but in tumor-infiltrating myeloid cells and that macrophage phagocytosis is inhibited in PD-L1-positive macrophages [26]. TIM expression of PD-L1 increases following dendritic cell vaccination, suggesting an evolving response to antitumor immune responses generated by the vaccine [27]. IL-6 has been implicated in GBM induction of PD-L1 in tumor-infiltrating myeloid cells [28]. Suppression of PD-L1 expression on tumor-infiltrating myeloid cells, either through the inhibition of IL-6 signaling [28] or through direct targeting of PD-L1 expression on these cells using nanoparticles [29], has been shown to inhibit tumor growth in mouse models. Further local immunosuppression is driven by intratumoral IL-10, which reduces major histocompatibility complex (MHC) class II cross-presentation in monocytes and prevents these cells from stimulating an immune response via antigen presentation [30]. MDSC populations therefore represent an attractive therapeutic target in GBM, supported by both preclinical and early-phase clinical trial data showing that low-dose metronomic capecitabine reduces circulating and tumor-associated MDSCs in recurrent GBM [31]. Other investigators have focused on enhancing myeloid lineage-mediated cellular antigen presentation in the TME; for example, the activation of dendritic cell-mediated antigen presentation was shown to improve response rates to checkpoint blockade in a mouse model of GBM [32].

TAMs, particularly M2-polarized macrophages/microglia, are key mediators of immunosuppression in GBM via cytokine signaling [33, 34], stimulation of T-cell apoptosis by expression of the Fas ligand [35], destruction of tumor-specific TILs leading to the induction of Tregs [36, 37], and PD-L1 expression [26]. In the physiologic state, M2 cells are major players in the resolution of brain parenchymal inflammation, and it is likely that tumor-induced inflammation contributes to the activation of M2-polarized TAMs within the TME in GBM [38].

Tumor-infiltrating lymphocytes are the third major cellular component of the immune microenvironment in GBM and again reflect the general landscape of immunosuppression seen in this disease. Compared with lower-grade gliomas, TIL populations in GBM are more likely to contain Tregs and have a higher CD4:CD8 ratio [39]. There is also a preponderance of antigen-exposed phenotype among the CD8+ TIL population in GBM, with reports differing over whether effector or central memory cells predominate [40, 41]. T-cell exhaustion, characterized by the upregulation of multiple immune checkpoints, is a prominent phenotype among cytotoxic CD8+ lymphocytes extracted from newly diagnosed GBM specimens, which commonly exhibit upregulation of LAG-3, TIGIT, and CD39 in addition to PD-1 [41]. An additional immune escape mechanism in GBM involves a shift in natural killer cell ligands from being expressed as cell surface markers to being released as soluble proteins [42]. This process is thought to "distract" cytotoxic effector cells as well as reducing the immunogenicity of the tumor cell itself [43].

Alterations in circulating and intratumoral immune cell populations are in part driven by direct tumor secretion of soluble immunosuppressive factors, such as the Treg chemoattractant CCL-2 [14], the monocyte chemoattractant MCP-1, which recruits tumor-associated macrophages to the TME [44], and immunosuppressive cytokines such as IL-6 [28] and IL-10 [45]. In addition to soluble factors, GBMs alter the immune phenotype by direct interaction between cancer cell surface markers and circulating immune cells. Perhaps the best studied of these interactions is the PD-L1/PD-1 axis, in which tumor cell expression of PD-L1 suppresses T-cell proliferation and activity via interactions with lymphocytic PD-1. PD-L1 expression in GBM appears to be a negative prognostic factor, but is highly variable, with between 10% and over 80% of GBM specimens demonstrating at least some detectable PD-L1 expression [46–49]. PD-1 induction on tumor-infiltrating T lymphocytes also appears to play a role in the induction of tumor-infiltrating myeloid-derived suppressor cells [50].

Indoleamine 2,3-dioxygenase (IDO) is another fundamental immunoregulatory pathway implicated in GBM-induced immunosuppression. Physiologically, macrophage expression of IDO is a critical regulator of maternal immune tolerance to the fetus; it exerts its action by depleting tryptophan, thereby inhibiting T-cell proliferation [51]. More recently, GBM-expressed IDO has also been shown to increase the ratio of Tregs to cytotoxic T cells within GBM-infiltrating lymphocyte populations, further enhancing the shift to Treg populations that is characteristic of the TME in GBM [52]. Over 90% of GBMs express IDO, and high levels are correlated with worse survival [53]. IDO expression within GBM appears to be mediated by T-cell infiltration and secretion of IFN-γ, consistent with other models that have reported that inflammatory cytokines induce IDO [54, 55].

STAT3, an oncogene involved in multiple critical biologic processes, including aerobic metabolism, inflammatory signaling, and embryogenesis, has been implicated as a key mediator of GBM-induced immunosuppression [56]. Constitutive activation of STAT3 is observed in up to 90% of GBMs, and STAT3 expression levels are associated with worse overall survival in multiple cancer types, including glioma [57]. Hypoxia within the tumor enhances STAT3 production in a HIF-1α-dependent manner [58]. In turn, STAT3 activation attracts monocytes from the circulation [59], modulates dendritic cell-based stimulation of Treg proliferation [60], and promotes the accumulation of immature myeloid cells by inhibiting differentiation into dendritic cells [61], ultimately enriching the glioma TME for immunosuppressive TAMs, MDSCs, and Tregs [28]. Additional downstream effects of tumor hypoxia on the TME include stimulation of the proangiogenic cytokines VEGF and EGF. VEGF has multiple effects on cellular immunity; it inhibits dendritic cell maturation [62], can directly stimulate Treg proliferation [63], and has been implicated in T-cell exhaustion [64]. Dual VEGF/Ang-2 inhibition has been reported to permit the transformation of immunosuppressive M2 macrophages back to an M1 phenotype in a mouse model of GBM [65] and to recruit MDSCs into tumor from the circulation. MDSCs are also an important source of VEGF, perpetuating a proangiogenic, proinvasive, and immunosuppressive milieu even after the original hypoxic signaling mechanism has faded [66].

Finally, intracranial location itself appears to be associated with differential effects on the immune response, as demonstrated in models of both brain metastases and primary brain tumors. For example, in a mouse model of metastatic melanoma, brain metastases were associated with significant impairments in antigen-specific cytotoxicity as well as elevated serum TGF-β compared with either flank or lung metastatic implants of the same cell line [67]. Similarly, in a mouse model of primary brain tumors, flank implants were associated with lower populations of tumor-infiltrating regulatory T cells; while both intracranial and flank tumors demonstrated infiltration with immature dendritic cells, more robust dendritic cell-mediated T-cell proliferation was observed among flank tumors than in primary intracranial tumors [68]. Intracranial tumors also appear to preferentially induce bone marrow T-cell sequestration compared with subcutaneous implants derived from either primary brain tumors or metastatic solid tumors, including breast cancer, lung cancer, and melanoma [20].

Treatment effects: Surgery

Surgical resection has been demonstrated to partially restore the function of circulating immune cells in GBM, likely by reducing the quantity of available tumor-supplied immunosuppressive factors [69]. However, surgery itself elicits a systemic stress response that can alter immune status through endogenous cortisol signaling [70]. Unsurprisingly, acute changes in circulating immune cell subtypes after intracranial surgery reflect dexamethasone-induced effects and include T-cell lymphocytopenia as well as anergic skin response to recall antigens [71]. Direct injury to the brain parenchyma during surgery, as well as a catecholamine-triggered immune response, has been implicated in the release of immunosuppressive cytokines, including IL-6 and IL-10 [72, 73]. While the postoperative serum elevation in IL-10 is transient (<24h), downstream effects of IL-10 signaling such as the inhibition of monocyte HLA-DR expression may persist for days, potentially reinforcing tumor-associated mechanisms of immunosuppression [74].

Anesthetic agents must cross the blood-brain barrier to exert their action, prompting investigations into potential antitumor effects of anesthetics in GBM. Propofol has received particular attention due to preclinical data, suggesting that it may enhance antitumor cytotoxic T-cell activity [75]. While some [76] investigators have suggested that total IV anesthesia may have preferential antitumor effects compared with volatile anesthetic agents, clinical data regarding the effect of anesthesia regimen on outcomes in GBM has been inconsistent. A relatively large multiinstitutional cohort from Germany reported that choice of anesthetic agent (volatile vs total IV anesthesia) had no effect on outcomes in a large multiinstitutional cohort of patients undergoing resection of primary brain tumors [77]. On the other hand, retrospective data suggest that regional anesthesia with scalp block for GBM resection is associated with lower postoperative levels of circulating inflammatory markers (neutrophil:lymphocyte and platelet:lymphocyte ratio), as well as longer progression-free survival [78].

Treatment effects: Corticosteroid therapy

GBM-induced alterations in the blood-brain barrier frequently cause severe peritumoral vasogenic edema. The corticosteroid medication dexamethasone is the mainstay of treatment for symptomatic vasogenic edema despite a constellation of systemic side effects associated with long-term use, including proximal myopathy, cushingoid appearance, hyperglycemia, bone loss, insomnia, emotional lability, gastric ulcers, and skin changes. Because nonsteroidal antiinflammatory drugs have a minimal effect on intracranial edema, dexamethasone is believed to exert its action partly via direct suppression of proinflammatory signals such as NF-κB [79]. Dexamethasone also inhibits VEGF and matrix metalloproteinases while upregulating proteins such as occludin and claudin, thus helping to restore the integrity of the blood-brain barrier [80].

Dexamethasone treatment is associated with frequent and sometimes dramatic leukocytosis and neutrophilia with a concomitant decrease in circulating eosinophils, monocytes, and lymphocytes [81]. Steroid-induced lymphopenia is characterized by relative T-cell depletion as well as a diminished response to antigen stimulation [82]. T-cell depletion is believed to be due in part to a redistribution of these cells out of the circulation [83]. Corticosteroid treatment also causes apoptosis of T lymphocytes [84] and inhibits the proliferation of naïve T cells by upregulating CTLA-4 [85]. While steroid-induced lymphopenia appears to be transient following single doses of steroids in normal controls [86], the prolonged use of steroids to control chronic intracranial edema causes chronic alterations in systemic immune status. Treatment with dexamethasone suppresses MHC class II expression on tumor-associated macrophages, suggesting that long-term steroid use may enhance TAM-induced immunosuppression within the tumor microenvironment [87]. Recently, it was reported that patients undergoing resection for GBM who developed dexamethasone-induced leukocytosis (defined as total white blood cell count above institutional upper limit of normal) had worse overall and progression-free survival than those whose white blood cell counts remained within the normal range and that dexamethasone-induced leukocytosis was associated with sparser CD3+ tumor-infiltrating lymphocyte populations, a finding also supported by animal brain tumor models [88, 89]. Suppression of T-cell infiltration into the TME by dexamethasone may be partially responsible for the inhibition of tumor-specific neoantigen generation in the setting of personalized antitumor vaccination, as reported in the results of a recent phase I study [90]. Supporting this hypothesis, post hoc analysis of Checkmate-143, which randomized patients with recurrent GBM to nivolumab vs bevacizumab, showed that MGMT-methylated patients who were not steroid-dependent at baseline were more likely to respond to checkpoint inhibition than steroid-dependent patients, although no effect of steroid use on outcomes with nivolumab was observed in MGMT-unmethylated patients [1]. Because dexamethasone is rarely administered as a stand-alone treatment in GBM (except in the palliative setting), it can be challenging to differentiate the immune effects of steroids from recurrent tumor as well as from other immunosuppressive interventions, especially radiotherapy and chemotherapy. However, in a sample of patients who were all treated with a uniform standard regimen of 60 Gy radiotherapy with concurrent temozolomide chemotherapy, patients who received dexamethasone had lower circulating Treg populations, as well as the expected steroid-induced "left shift" in circulating white blood cell populations (i.e., enrichment for neutrophils and a lower percentage of lymphocytes). In all patients, regardless of steroid use, CD4:CD8 ratio decreased significantly with disease progression [91].

Treatment effects: Chemotherapy and radiotherapy

Because concurrent radiotherapy and temozolomide are routinely administered to all patients receiving definitive therapy for GBM as part of the standard of care, it can be challenging to discern which of these potentially immunotoxic therapies are responsible for individual alterations in immune status among patients undergoing definitive therapy for GBM. Despite these difficulties, observations from prior studies of radiation and chemotherapy alone, as well as in combination, can be used to inform our conclusions. The risk of radiation-induced lymphopenia is not unique to patients with brain tumors and has been observed following focal fractionated radiotherapy to all anatomic sites, with rates of severe radiation-induced lymphopenia approaching 75% in certain diseases such as esophageal cancer [92]. Treatment-induced lymphopenia in patients undergoing focal fractionated radiation is due to direct toxicity to circulating immune cells that pass through the irradiated field while the beam is on [93]; radiation-induced cell killing is likely enhanced, rather than caused, by the concurrent administration of lymphotoxic chemotherapy agents such as temozolomide [92].

Support for the irradiation of circulating immune cells as the main driver of lymphopenia among patients undergoing partial-body radiation, including radiation for brain tumors, is provided by a wealth of prior preclinical and clinical data. Although the skull contains a small amount of bone marrow, and microscopic intracranial lymphatic vessels have recently been identified in humans, there are no other primary or secondary lymphoid structures within either the brain itself or the cranium. Circulating lymphocytes are therefore the key target tissue for lymphocyte toxicity during brain irradiation, consistent with the fact that the brain receives approximately 15% of resting cardiac output [94]. Further support for this hypothesis is derived from clinical data, including observations that irradiation of circulating blood alone via a radioactive source placed within a shielded dialysis unit leads to profound and prolonged depletion of circulating lymphocyte counts, resulting in sufficient immunosuppression to prevent solid organ allograft rejection, as well as multiple observational studies demonstrating the ability of focal fractionated radiation alone to induce significant decreases in peripheral lymphocyte counts [95, 96].

Concurrent chemoradiation for primary brain tumors causes characteristic patterns of systemic immunosuppression and changes in the tumor immune microenvironment. Even without concurrent temozolomide, opportunistic infections such as *Pneumocystis* pneumonia are frequently reported in patients undergoing radiation for primary brain tumors. Most patients have a history of prolonged steroid use [97]. In addition to *Pneumocystis*, disseminated strongyloidiasis [98], cryptococcal meningitis [99], and other opportunistic infections have been reported during definitive therapy for GBM [100]. Subsequent analyses revealed that severe CD4+ lymphopenia is a common side effect of radiotherapy and corticosteroid use, with 24% of patients treated with this regimen developing CD4+ cell counts <200 cells/mm^3 [101]. The addition of temozolomide appears to further potentiate treatment-induced CD4+ lymphopenia, with 40% of patients developing CD4+ count <200 cells/mm^3 in the only large prospective study to track changes in CD4+ count during concurrent temozolomide and radiotherapy for malignant glioma [13]. Treatment-induced lymphopenia is a clear negative prognostic factor in GBM, with several studies confirming a nearly twofold increase in the hazard ratio for death among patients who develop grade 3 or greater lymphopenia following chemoradiation for high-grade glioma [13, 102, 103]. In GBM patients with treatment-induced lymphopenia, death is predominantly due to tumor progression and not opportunistic infection, a particularly intriguing finding in a tumor that is widely considered to be immunologically "cold." Strategies to overcome radiation-induced immunosuppression in GBM include the use of hypofractionated radiation regimens, which reduce the exposure of circulating immune cells to potentially damaging radiation doses, as well as interventions such as autologous T-cell transfer and IL-7 administration, which may augment lymphocyte recovery.

The mainstay of cytotoxic chemotherapy for GBM is the alkylating agent temozolomide. Second-line drugs approved for use in GBM include the VEGF inhibitor bevacizumab, the legacy alkylating agents BCNU (lomustine) and CCNU (carmustine), as well as etoposide, which inhibits DNA topoisomerase II, and regorafenib, which is a multitargeted tyrosine kinase inhibitor. Temozolomide causes dose-dependent peripheral lymphopenia and is associated with selective toxicity to CD4+ cells in particular. Patients with temozolomide-induced lymphopenia are clinically immunosuppressed, with an increased risk of opportunistic infections [104]. Temozolomide's effects on circulating lymphocyte counts are dose- and schedule-dependent; in a comparison of dose-dense (100 mg/m^2 on 21/28 days) versus standard-dose (200 mg/m^2 on 5/28 days) temozolomide among patients with recurrent GBM undergoing EGFR-targeted antitumor vaccination, the dose-dense schedule was associated with uniform induction of grade 3 or greater lymphopenia by the sixth cycle of treatment, while patients treated with the standard-dose regimen had only a 16% rate of $\geq$grade 3 lymphopenia at the corresponding time point [105]. Dose-dense treatment was also associated with greater depletion of total B and T cells as well as a significant increase in circulating Tregs proportion. Despite the increased Treg fraction, antibody response to anti-EGFR vaccine was greatest in the dose-dense group. The investigators postulated that temozolomide-induced lymphopenia may actually enhance immune responses by favoring the proliferation of T cells that were primed by the vaccine to respond to tumor antigen during the period of recovery from temozolomide-induced lymphopenia (vaccine was administered following the completion of postoperative chemoradiation but before beginning adjuvant single-agent temozolomide).

Characteristic changes in circulating lymphocyte subtypes following temozolomide-based chemoradiation for GBM include decreased circulating B cells (CD19+), total T cell (CD3+), and CD4+ T cells, with less pronounced effects on CD8+ and Treg cells, leading to concomitant decreases in the CD4:CD8 ratio and increases in the proportion of circulating Tregs [106–108]. In patients with treatment-induced lymphopenia, the expected compensatory increase in IL-7 and IL-15 in response to lymphopenia does not appear to occur, although the mechanism responsible for dysfunctional cytokine responses in these patients remains unclear [107]. Exogenous IL-7 is under active investigation as a potential strategy for restoring circulating lymphocyte populations following chemoradiotherapy for GBM [109].

The tumor microenvironment is also modified by standard therapies for GBM. Some investigators have proposed that radiation can enhance antitumor immune responses by upregulating local inflammatory responses and stimulating release of tumor neoantigens, thus acting as an "*in situ* vaccine." [110] However, chronic inflammation and hypoxia induced by radiotherapy can also result in local immunosuppression, which is likely mediated by a combination of T-cell exhaustion and the induction of stromal proliferation leading to tissue fibrosis [111, 112]. The balance of such effects is strongly dependent on dose and fractionation schedule [113]. Radiation-induced hypoxia is believed to promote local immunosuppression through the promotion of HIF-1- and CXCL12-dependent pathways that are involved in MDSC recruitment into the TME [114]. Temozolomide and the older alkylating agent carmustine reduce CCL-2 secretion by glioma cells, thereby potentially decreasing intratumoral Treg and MDSC populations [115]. Preclinical data also suggest that temozolomide may enhance antitumor effects of CCL-2 blockade [116]. Unfortunately, to date, only limited data is available on the ability of monotherapy with alkylating agents to alter the composition of the immune cell infiltrate within the TME.

One of the most potent strategies for immune modulation in the TME is VEGF blockade, given the manifold immunosuppressive effects of tumor VEGF secretion. As described in detail above, VEGF has multiple immunosuppressive effects within the TME, including MDSC recruitment, inhibition of dendritic cell maturation, and M2 polarization of tumor-associated macrophages. Despite potentially promising synergy between checkpoint blockade-based immunotherapies and VEGF inhibition, as well as positive clinical trial results for combinations of immune checkpoint blockade and anti-VEGF agents in other diseases such as hepatocellular carcinoma [117], early-phase clinical trials in recurrent GBM have not demonstrated any benefit to combined therapy with bevacizumab and immune checkpoint inhibitors [118, 119]. Results of additional ongoing studies of ICI/anti-VEGF combinations are awaited.

Conclusions

To date, efforts to harness antitumor immune responses against GBM have been confounded by the profound immunosuppression characteristic of this disease, which is induced by both the tumor itself and its treatments. Multipronged strategies that capitalize on our increasingly sophisticated understanding of the pathophysiology of immunosuppression in GBM will likely be needed to overcome treatment resistance and improve outcomes.

References

[1] Reardon DA, Brandes AA, Omuro A, Mulholland P, Lim M, Wick A, et al. Effect of nivolumab vs bevacizumab in patients with recurrent glioblastoma: the CheckMate 143 phase 3 randomized clinical trial. JAMA Oncol 2020.

[2] Anon. Bristol-Myers Squibb announces phase 3 CheckMate-498 study did not met primary endpoint of overall survival with Opdivo (nivolumab) plus radiation in patients with newly diagnosed MGMT-unmethylated glioblastoma multiforme. Bristol-Myers Squibb; 2019. [updated May 9, 2019; cited July 2, 2020]. Available from: https://news.bms.com/press-release/corporatefinancial-news/bristol-myers-squibb-announces-phase-3-checkmate-498-study-did.

[3] Anon. Bristol-Myers Squibb provides update on phase 3 Opdivo (nivolumab) CheckMate-548 trial in patients with newly diagnosed MGMT-methylated glioblastoma multiforme. Bristol-Myers Squibb; 2019. [updated September 5, 2019; cited July 2, 2020]. Available from: https://news.bms.com/press-release/corporatefinancial-news/bristol-myers-squibb-provides-update-phase-3-opdivo-nivolumab-.

[4] Kida S, Pantazis A, Weller RO. CSF drains directly from the subarachnoid space into nasal lymphatics in the rat. Anatomy, histology and immunological significance. Neuropathol Appl Neurobiol 1993;19(6):480–8.

[5] Aspelund A, Antila S, Proulx ST, Karlsen TV, Karaman S, Detmar M, et al. A dural lymphatic vascular system that drains brain interstitial fluid and macromolecules. J Exp Med 2015;212(7):991–9.

[6] Louveau A, Smirnov I, Keyes TJ, Eccles JD, Rouhani SJ, Peske JD, et al. Structural and functional features of central nervous system lymphatic vessels. Nature 2015;523(7560):337–41.

[7] Absinta M, Ha SK, Nair G, Sati P, Luciano NJ, Palisoc M, et al. Human and nonhuman primate meninges harbor lymphatic vessels that can be visualized noninvasively by MRI. elife 2017;6. https://doi.org/10.7554/eLife.29738.

[8] Russo MV, McGavern DB. Immune surveillance of the CNS following infection and injury. Trends Immunol 2015;36(10):637–50.

[9] Lauterbach H, Zuniga EI, Truong P, Oldstone MB, McGavern DB. Adoptive immunotherapy induces CNS dendritic cell recruitment and antigen presentation during clearance of a persistent viral infection. J Exp Med 2006;203(8):1963–75.

[10] Brooks WH, Netsky MG, Normansell DE, Horwitz DA. Depressed cell-mediated immunity in patients with primary intracranial tumors. Characterization of a humoral immunosuppressive factor. J Exp Med 1972;136(6):1631–47.

[11] Dix AR, Brooks WH, Roszman TL, Morford LA. Immune defects observed in patients with primary malignant brain tumors. J Neuroimmunol 1999;100(1–2):216–32.

[12] Bhondeley MK, Mehra RD, Mehra NK, Mohapatra AK, Tandon PN, Roy S, et al. Imbalances in T cell subpopulations in human gliomas. J Neurosurg 1988;68(4):589–93.

[13] Grossman SA, Ye X, Lesser G, Sloan A, Carraway H, Desideri S, et al. Immunosuppression in patients with high-grade gliomas treated with radiation and temozolomide. Clin Cancer Res 2011;17(16):5473–80.

[14] Crane CA, Ahn BJ, Han SJ, Parsa AT. Soluble factors secreted by glioblastoma cell lines facilitate recruitment, survival, and expansion of regulatory T cells: implications for immunotherapy. Neuro-Oncology 2012;14(5):584–95.

[15] Fecci PE, Mitchell DA, Whitesides JF, Xie W, Friedman AH, Archer GE, et al. Increased regulatory T-cell fraction amidst a diminished CD4 compartment explains cellular immune defects in patients with malignant glioma. Cancer Res 2006;66(6):3294–302.

[16] Dieckmann D, Plottner H, Berchtold S, Berger T, Schuler G. Ex vivo isolation and characterization of CD4(+) CD25(+) T cells with regulatory properties from human blood. J Exp Med 2001;193(11):1303–10.

[17] Ashkenazi E, Deutsch M, Tirosh R, Weinreb A, Tsukerman A, Brodie C. A selective impairment of the IL-2 system in lymphocytes of patients with glioblastomas: increased level of soluble IL-2R and reduced protein tyrosine phosphorylation. Neuroimmunomodulation 1997;4(1):49–56.

[18] Morford LA, Elliott LH, Carlson SL, Brooks WH, Roszman TL. T cell receptor-mediated signaling is defective in T cells obtained from patients with primary intracranial tumors. J Immunol 1997;159(9):4415–25.
[19] Fornara O, Odeberg J, Wolmer Solberg N, Tammik C, Skarman P, Peredo I, et al. Poor survival in glioblastoma patients is associated with early signs of immunosenescence in the CD4 T-cell compartment after surgery. Onco Targets Ther 2015;4(9), e1036211.
[20] Chongsathidkiet P, Jackson C, Koyama S, Loebel F, Cui X, Farber SH, et al. Sequestration of T cells in bone marrow in the setting of glioblastoma and other intracranial tumors. Nat Med 2018;24(9):1459–68.
[21] Raychaudhuri B, Rayman P, Ireland J, Ko J, Rini B, Borden EC, et al. Myeloid-derived suppressor cell accumulation and function in patients with newly diagnosed glioblastoma. Neuro-Oncology 2011;13 (6):591–9.
[22] Alban TJ, Alvarado AG, Sorensen MD, Bayik D, Volovetz J, Serbinowski E, et al. Global immune fingerprinting in glioblastoma patient peripheral blood reveals immune-suppression signatures associated with prognosis. JCI Insight 2018;3(21). https://doi.org/10.1172/jci.insight.122264.
[23] Baumann T, Dunkel A, Schmid C, Schmitt S, Hiltensperger M, Lohr K, et al. Regulatory myeloid cells paralyze T cells through cell-cell transfer of the metabolite methylglyoxal. Nat Immunol 2020;21(5):555–66.
[24] Thorsson V, Gibbs DL, Brown SD, Wolf D, Bortone DS, Ou Yang TH, et al. The immune landscape of cancer. Immunity 2019;51(2):411–2.
[25] Amankulor NM, Kim Y, Arora S, Kargl J, Szulzewsky F, Hanke M, et al. Mutant IDH1 regulates the tumor-associated immune system in gliomas. Genes Dev 2017;31(8):774–86.
[26] Gordon SR, Maute RL, Dulken BW, Hutter G, George BM, McCracken MN, et al. PD-1 expression by tumour-associated macrophages inhibits phagocytosis and tumour immunity. Nature 2017;545(7655):495–9.
[27] Antonios JP, Soto H, Everson RG, Moughon D, Orpilla JR, Shin NP, et al. Immunosuppressive tumor-infiltrating myeloid cells mediate adaptive immune resistance via a PD-1/PD-L1 mechanism in glioblastoma. Neuro-Oncology 2017;19(6):796–807.
[28] Lamano JB, Lamano JB, Li YD, DiDomenico JD, Choy W, Veliceasa D, et al. Glioblastoma-derived IL6 induces immunosuppressive peripheral myeloid cell PD-L1 and promotes tumor growth. Clin Cancer Res 2019;25 (12):3643–57.
[29] Zhang P, Miska J, Lee-Chang C, Rashidi A, Panek WK, An S, et al. Therapeutic targeting of tumor-associated myeloid cells synergizes with radiation therapy for glioblastoma. Proc Natl Acad Sci U S A 2019;116(47): 23714–23.
[30] Mittal SK, Roche PA. Suppression of antigen presentation by IL-10. Curr Opin Immunol 2015;34:22–7.
[31] Peereboom DM, Alban TJ, Grabowski MM, Alvarado AG, Otvos B, Bayik D, et al. Metronomic capecitabine as an immune modulator in glioblastoma patients reduces myeloid-derived suppressor cells. JCI Insight 2019;4 (22). https://doi.org/10.1172/jci.insight.130748.
[32] Garzon-Muvdi T, Theodros D, Luksik AS, Maxwell R, Kim E, Jackson CM, et al. Dendritic cell activation enhances anti-PD-1 mediated immunotherapy against glioblastoma. Oncotarget 2018;9(29):20681–97.
[33] Huettner C, Czub S, Kerkau S, Roggendorf W, Tonn JC. Interleukin 10 is expressed in human gliomas in vivo and increases glioma cell proliferation and motility in vitro. Anticancer Res 1997;17(5A):3217–24.
[34] Zhang J, Sarkar S, Cua R, Zhou Y, Hader W, Yong VW. A dialog between glioma and microglia that promotes tumor invasiveness through the CCL2/CCR2/interleukin-6 axis. Carcinogenesis 2012;33(2):312–9.
[35] Badie B, Schartner J, Prabakaran S, Paul J, Vorpahl J. Expression of Fas ligand by microglia: possible role in glioma immune evasion. J Neuroimmunol 2001;120(1–2):19–24.
[36] Xu L, Xiao H, Xu M, Zhou C, Yi L, Liang H. Glioma-derived T cell immunoglobulin- and mucin domain-containing molecule-4 (TIM4) contributes to tumor tolerance. J Biol Chem 2011;286(42):36694–9.
[37] Ebner F, Brandt C, Thiele P, Richter D, Schliesser U, Siffrin V, et al. Microglial activation milieu controls regulatory T cell responses. J Immunol 2013;191(11):5594–602.
[38] Cherry JD, Olschowka JA, O'Banion MK. Neuroinflammation and M2 microglia: the good, the bad, and the inflamed. J Neuroinflammation 2014;11:98. https://doi.org/10.1186/1742-2094-11-98.
[39] Han S, Zhang C, Li Q, Dong J, Liu Y, Huang Y, et al. Tumour-infiltrating CD4(+) and CD8(+) lymphocytes as predictors of clinical outcome in glioma. Br J Cancer 2014;110(10):2560–8.
[40] Liu Z, Meng Q, Bartek Jr J, Poiret T, Persson O, Rane L, et al. Tumor-infiltrating lymphocytes (TILs) from patients with glioma. Onco Targets Ther 2016;6(2), e1252894.
[41] Woroniecka K, Chongsathidkiet P, Rhodin K, Kemeny H, Dechant C, Farber SH, et al. T-cell exhaustion signatures vary with tumor type and are severe in glioblastoma. Clin Cancer Res 2018;24(17):4175–86.

[42] Chitadze G, Lettau M, Luecke S, Wang T, Janssen O, Furst D, et al. NKG2D- and T-cell receptor-dependent lysis of malignant glioma cell lines by human gammadelta T cells: modulation by temozolomide and a disintegrin and metalloproteases 10 and 17 inhibitors. Onco Targets Ther 2015;5(4), e1093276.
[43] Salih HR, Holdenrieder S, Steinle A. Soluble NKG2D ligands: prevalence, release, and functional impact. Front Biosci 2008;13:3448–56.
[44] Desbaillets I, Tada M, de Tribolet N, Diserens AC, Hamou MF, Van Meir EG. Human astrocytomas and glioblastomas express monocyte chemoattractant protein-1 (MCP-1) in vivo and in vitro. Int J Cancer 1994;58 (2):240–7.
[45] Bloch O, Crane CA, Kaur R, Safaee M, Rutkowski MJ, Parsa AT. Gliomas promote immunosuppression through induction of B7-H1 expression in tumor-associated macrophages. Clin Cancer Res 2013;19(12):3165–75.
[46] Nduom EK, Wei J, Yaghi NK, Huang N, Kong LY, Gabrusiewicz K, et al. PD-L1 expression and prognostic impact in glioblastoma. Neuro-Oncology 2016;18(2):195–205.
[47] Hodges TR, Ott M, Xiu J, Gatalica Z, Swensen J, Zhou S, et al. Mutational burden, immune checkpoint expression, and mismatch repair in glioma: implications for immune checkpoint immunotherapy. Neuro-Oncology 2017;19(8):1047–57.
[48] Berghoff AS, Kiesel B, Widhalm G, Wilhelm D, Rajky O, Kurscheid S, et al. Correlation of immune phenotype with IDH mutation in diffuse glioma. Neuro-Oncology 2017;19(11):1460–8.
[49] Chen RQ, Liu F, Qiu XY, Chen XQ. The prognostic and therapeutic value of PD-L1 in glioma. Front Pharmacol 2019;9:1503.
[50] Dubinski D, Wolfer J, Hasselblatt M, Schneider-Hohendorf T, Bogdahn U, Stummer W, et al. CD4+ T effector memory cell dysfunction is associated with the accumulation of granulocytic myeloid-derived suppressor cells in glioblastoma patients. Neuro-Oncology 2016;18(6):807–18.
[51] Munn DH, Shafizadeh E, Attwood JT, Bondarev I, Pashine A, Mellor AL. Inhibition of T cell proliferation by macrophage tryptophan catabolism. J Exp Med 1999;189(9):1363–72.
[52] Wainwright DA, Balyasnikova IV, Chang AL, Ahmed AU, Moon KS, Auffinger B, et al. IDO expression in brain tumors increases the recruitment of regulatory T cells and negatively impacts survival. Clin Cancer Res 2012;18 (22):6110–21.
[53] Mitsuka K, Kawataki T, Satoh E, Asahara T, Horikoshi T, Kinouchi H. Expression of indoleamine 2,3-dioxygenase and correlation with pathological malignancy in gliomas. Neurosurgery 2013;72(6):1031–8 [discussion 1038-9].
[54] Zhai L, Ladomersky E, Lauing KL, Wu M, Genet M, Gritsina G, et al. Infiltrating T cells increase IDO1 expression in glioblastoma and contribute to decreased patient survival. Clin Cancer Res 2017;23(21):6650–60.
[55] Carlin JM, Borden EC, Sondel PM, Byrne GI. Biologic-response-modifier-induced indoleamine 2,3-dioxygenase activity in human peripheral blood mononuclear cell cultures. J Immunol 1987;139(7):2414–8.
[56] Avalle L, Camporeale A, Camperi A, Poli V. STAT3 in cancer: a double edged sword. Cytokine 2017;98:42–50.
[57] Wu P, Wu D, Zhao L, Huang L, Shen G, Huang J, et al. Prognostic role of STAT3 in solid tumors: a systematic review and meta-analysis. Oncotarget 2016;7(15):19863–83.
[58] Niu G, Briggs J, Deng J, Ma Y, Lee H, Kortylewski M, et al. Signal transducer and activator of transcription 3 is required for hypoxia-inducible factor-1alpha RNA expression in both tumor cells and tumor-associated myeloid cells. Mol Cancer Res 2008;6(7):1099–105.
[59] Liu YS, Lin HY, Lai SW, Huang CY, Huang BR, Chen PY, et al. MiR-181b modulates EGFR-dependent VCAM-1 expression and monocyte adhesion in glioblastoma. Oncogene 2017;36(35):5006–22.
[60] Matsumura Y, Kobayashi T, Ichiyama K, Yoshida R, Hashimoto M, Takimoto T, et al. Selective expansion of foxp3-positive regulatory T cells and immunosuppression by suppressors of cytokine signaling 3-deficient dendritic cells. J Immunol 2007;179(4):2170–9.
[61] Nefedova Y, Huang M, Kusmartsev S, Bhattacharya R, Cheng P, Salup R, et al. Hyperactivation of STAT3 is involved in abnormal differentiation of dendritic cells in cancer. J Immunol 2004;172(1):464–74.
[62] Gabrilovich DI, Chen HL, Girgis KR, Cunningham HT, Meny GM, Nadaf S, et al. Production of vascular endothelial growth factor by human tumors inhibits the functional maturation of dendritic cells. Nat Med 1996;2 (10):1096–103.
[63] Terme M, Pernot S, Marcheteau E, Sandoval F, Benhamouda N, Colussi O, et al. VEGFA-VEGFR pathway blockade inhibits tumor-induced regulatory T-cell proliferation in colorectal cancer. Cancer Res 2013;73 (2):539–49.

[64] Voron T, Colussi O, Marcheteau E, Pernot S, Nizard M, Pointet AL, et al. VEGF-A modulates expression of inhibitory checkpoints on CD8+ T cells in tumors. J Exp Med 2015;212(2):139–48.
[65] Kloepper J, Riedemann L, Amoozgar Z, Seano G, Susek K, Yu V, et al. Ang-2/VEGF bispecific antibody reprograms macrophages and resident microglia to anti-tumor phenotype and prolongs glioblastoma survival. Proc Natl Acad Sci U S A 2016;113(16):4476–81.
[66] Berse B, Brown LF, Van de Water L, Dvorak HF, Senger DR. Vascular permeability factor (vascular endothelial growth factor) gene is expressed differentially in normal tissues, macrophages, and tumors. Mol Biol Cell 1992;3 (2):211–20.
[67] Jackson CM, Kochel CM, Nirschl CJ, Durham NM, Ruzevick J, Alme A, et al. Systemic tolerance mediated by melanoma brain tumors is reversible by radiotherapy and vaccination. Clin Cancer Res 2016;22(5):1161–72.
[68] Biollaz G, Bernasconi L, Cretton C, Puntener U, Frei K, Fontana A, et al. Site-specific anti-tumor immunity: differences in DC function, TGF-beta production and numbers of intratumoral Foxp3+ Treg. Eur J Immunol 2009;39(5):1323–33.
[69] Woiciechowsky C, Asadullah K, Nestler D, Schoning B, Glockner F, Docke WD, et al. Diminished monocytic HLA-DR expression and ex vivo cytokine secretion capacity in patients with glioblastoma: effect of tumor extirpation. J Neuroimmunol 1998;84(2):164–71.
[70] Alhayyan AM, McSorley ST, Kearns RJ, Horgan PG, Roxburgh CSD, McMillan DC. The relationship between anaesthetic technique, clinicopathological characteristics and the magnitude of the postoperative systemic inflammatory response in patients undergoing elective surgery for colon cancer. PLoS ONE 2020;15(4), e0228580.
[71] Dauch WA, Krex D, Heymanns J, Zeithammer B, Bauer BL. Peri-operative changes of cellular and humoral components of immunity with brain tumour surgery. Acta Neurochir 1994;126(2–4):93–101.
[72] Woiciechowsky C, Asadullah K, Nestler D, Glockner F, Robinson PN, Volk HD, et al. Different release of cytokines into the cerebrospinal fluid following surgery for intra- and extra-axial brain tumours. Acta Neurochir 1997;139(7):619–24.
[73] Woiciechowsky C, Asadullah K, Nestler D, Eberhardt B, Platzer C, Schoning B, et al. Sympathetic activation triggers systemic interleukin-10 release in immunodepression induced by brain injury. Nat Med 1998;4 (7):808–13.
[74] Klava A, Windsor AC, Farmery SM, Woodhouse LF, Reynolds JV, Ramsden CW, et al. Interleukin-10. A role in the development of postoperative immunosuppression. Arch Surg 1997;132(4):425–9.
[75] Kushida A, Inada T, Shingu K. Enhancement of antitumor immunity after propofol treatment in mice. Immunopharmacol Immunotoxicol 2007;29(3–4):477–86.
[76] Wigmore TJ, Mohammed K, Jhanji S. Long-term survival for patients undergoing volatile versus IV anesthesia for cancer surgery: a retrospective analysis. Anesthesiology 2016;124(1):69–79.
[77] Grau SJ, Lohr M, Taurisano V, Trautner H, Timmer M, Schwab SG, et al. The choice of anaesthesia for glioblastoma surgery does not impact the time to recurrence. Sci Rep 2020;10(1):5556. https://doi.org/10.1038/s41598-020-62087-8.
[78] Zheng L, Hagan KB, Villarreal J, Keerty V, Chen J, Cata JP. Scalp block for glioblastoma surgery is associated with lower inflammatory scores and improved survival. Minerva Anestesiol 2017;83(11):1137–45.
[79] Auphan N, DiDonato JA, Rosette C, Helmberg A, Karin M. Immunosuppression by glucocorticoids: inhibition of NF-kappa B activity through induction of I kappa B synthesis. Science 1995;270(5234):286–90.
[80] Murayi R, Chittiboina P. Glucocorticoids in the management of peritumoral brain edema: a review of molecular mechanisms. Childs Nerv Syst 2016;32(12):2293–302.
[81] Fauci AS, Dale DC, Balow JE. Glucocorticosteroid therapy: mechanisms of action and clinical considerations. Ann Intern Med 1976;84(3):304–15.
[82] Fauci AS, Dale DC. The effect of in vivo hydrocortisone on subpopulations of human lymphocytes. J Clin Invest 1974;53(1):240–6.
[83] Claman HN, Moorhead JW, Benner WH. Corticosteroids and lymphoid cells in vitro. I. Hydrocortisone lysis of human, Guinea pig, and mouse thymus cells. J Lab Clin Med 1971;78(4):499–507.
[84] Marchetti MC, Di Marco B, Cifone G, Migliorati G, Riccardi C. Dexamethasone-induced apoptosis of thymocytes: role of glucocorticoid receptor-associated Src kinase and caspase-8 activation. Blood 2003;101(2):585–93.
[85] Giles AJ, Hutchinson MND, Sonnemann HM, Jung J, Fecci PE, Ratnam NM, et al. Dexamethasone-induced immunosuppression: mechanisms and implications for immunotherapy. J Immunother Cancer 2018;6 (1):51. https://doi.org/10.1186/s40425-018-0371-5.

[86] Parrillo JE, Fauci AS. Mechanisms of corticosteroid action on lymphocyte subpopulations. III. Differential effects of dexamethasone administration on subpopulations of effector cells mediating cellular cytotoxicity in man. Clin Exp Immunol 1978;31(1):116–25.

[87] Moyes KW, Davis A, Hoglund V, Haberthur K, Lieberman NA, Kreuser SA, et al. Effects of tumor grade and dexamethasone on myeloid cells in patients with glioma. Onco Targets Ther 2018;7(11), e1507668.

[88] Dubinski D, Won SY, Gessler F, Quick-Weller J, Behmanesh B, Bernatz S, et al. Dexamethasone-induced leukocytosis is associated with poor survival in newly diagnosed glioblastoma. J Neuro-Oncol 2018;137(3):503–10.

[89] Badie B, Schartner JM, Paul J, Bartley BA, Vorpahl J, Preston JK. Dexamethasone-induced abolition of the inflammatory response in an experimental glioma model: a flow cytometry study. J Neurosurg 2000;93(4):634–9.

[90] Keskin DB, Anandappa AJ, Sun J, Tirosh I, Mathewson ND, Li S, et al. Neoantigen vaccine generates intratumoral T cell responses in phase Ib glioblastoma trial. Nature 2019;565(7738):234–9.

[91] Chitadze G, Fluh C, Quabius ES, Freitag-Wolf S, Peters C, Lettau M, et al. In-depth immunophenotyping of patients with glioblastoma multiforme: impact of steroid treatment. Onco Targets Ther 2017;6(11), e1358839.

[92] Ellsworth SG. Field size effects on the risk and severity of treatment-induced lymphopenia in patients undergoing radiation therapy for solid tumors. Adv Radiat Oncol 2018;3(4):512–9.

[93] Yovino S, Kleinberg L, Grossman SA, Narayanan M, Ford E. The etiology of treatment-related lymphopenia in patients with malignant gliomas: modeling radiation dose to circulating lymphocytes explains clinical observations and suggests methods of modifying the impact of radiation on immune cells. Cancer Investig 2013;31 (2):140–4.

[94] Ganong W, editor. Review of medical physiology. 21st ed. New York, NY: Lange Medical Books, McGraw-Hill; 2003.

[95] Weeke E. The development of lymphopenia in uremic patients undergoing extracorporeal irradiation of the blood with portable beta units. Radiat Res 1973;56(3):554–9.

[96] Grassberger C, Ellsworth SG, Wilks MQ, Keane FK, Loeffler JS. Assessing the interactions between radiotherapy and antitumour immunity. Nat Rev Clin Oncol 2019;16(12):729–45.

[97] Henson JW, Jalaj JK, Walker RW, Stover DE, Fels AO. Pneumocystis carinii pneumonia in patients with primary brain tumors. Arch Neurol 1991;48(4):406–9.

[98] Aregawi D, Lopez D, Wick M, Scheld WM, Schiff D. Disseminated strongyloidiasis complicating glioblastoma therapy: a case report. J Neuro-Oncol 2009;94(3):439–43.

[99] Choi JD, Powers CJ, Vredenburgh JJ, Friedman AH, Sampson JH. Cryptococcal meningitis in patients with glioma: a report of two cases. J Neuro-Oncol 2008;89(1):51–3.

[100] Munhoz RR, Pereira Picarelli AA, Troques Mitteldorf CA, Feher O. Aspergillosis in a patient receiving temozolomide for the treatment of glioblastoma. Case Rep Oncol 2013;6(2):410–5.

[101] Hughes MA, Parisi M, Grossman S, Kleinberg L. Primary brain tumors treated with steroids and radiotherapy: low CD4 counts and risk of infection. Int J Radiat Oncol Biol Phys 2005;62(5):1423–6.

[102] Mendez JS, Govindan A, Leong J, Gao F, Huang J, Campian JL. Association between treatment-related lymphopenia and overall survival in elderly patients with newly diagnosed glioblastoma. J Neuro-Oncol 2016;127(2):329–35.

[103] Rudra S, Hui C, Rao YJ, Samson P, Lin AJ, Chang X, et al. Effect of radiation treatment volume reduction on lymphopenia in patients receiving chemoradiotherapy for glioblastoma. Int J Radiat Oncol Biol Phys 2018;101 (1):217–25.

[104] Su YB, Sohn S, Krown SE, Livingston PO, Wolchok JD, Quinn C, et al. Selective CD4+ lymphopenia in melanoma patients treated with temozolomide: a toxicity with therapeutic implications. J Clin Oncol 2004;22(4): 610–6.

[105] Sampson JH, Aldape KD, Archer GE, Coan A, Desjardins A, Friedman AH, et al. Greater chemotherapy-induced lymphopenia enhances tumor-specific immune responses that eliminate EGFRvIII-expressing tumor cells in patients with glioblastoma. Neuro-Oncology 2011;13(3):324–33.

[106] Fadul CE, Fisher JL, Gui J, Hampton TH, Cote AL, Ernstoff MS. Immune modulation effects of concomitant temozolomide and radiation therapy on peripheral blood mononuclear cells in patients with glioblastoma multiforme. Neuro-Oncology 2011;13(4):393–400.

[107] Ellsworth S, Balmanoukian A, Kos F, Nirschl C, Nirschl T, Grossman S, et al. Sustained CD4+ cell driven lymphopenia without a compensatory IL7/Il15 response among high-grade glioma patients treated with radiation and temozolomide. Oncoimmunology 2014. e-pub ahead of print.

[108] Dutoit V, Philippin G, Widmer V, Marinari E, Vuilleumier A, Migliorini D, et al. Impact of radiochemotherapy on immune cell subtypes in high-grade glioma patients. Front Oncol 2020;10:89.
[109] Anon. Study of the effect of NT-I7 on CD4 counts in patients with high grade glioma. Bethesda, MD: U.S. National Library of Medicine; 2020. [updated July 2; cited July 2, 2020]. Available from: https://www.clinicaltrials.gov/ct2/show/NCT02659800?term=nt-i7&cond=glioma&draw=2&rank=1.
[110] Demaria S, Ng B, Devitt ML, Babb JS, Kawashima N, Liebes L, et al. Ionizing radiation inhibition of distant untreated tumors (abscopal effect) is immune mediated. Int J Radiat Oncol Biol Phys 2004;58(3):862–70.
[111] Dovedi SJ, Adlard AL, Lipowska-Bhalla G, McKenna C, Jones S, Cheadle EJ, et al. Acquired resistance to fractionated radiotherapy can be overcome by concurrent PD-L1 blockade. Cancer Res 2014;74(19):5458–68.
[112] Valkenburg KC, de Groot AE, Pienta KJ. Targeting the tumour stroma to improve cancer therapy. Nat Rev Clin Oncol 2018;15(6):366–81.
[113] Arnold KM, Flynn NJ, Raben A, Romak L, Yu Y, Dicker AP, et al. The impact of radiation on the tumor microenvironment: effect of dose and fractionation schedules. Cancer Growth Metastasis 2018;11, 1179064418761639.
[114] Thomas RP, Nagpal S, Iv M, Soltys SG, Bertrand S, Pelpola JS, et al. Macrophage exclusion after radiation therapy (MERT): a first in human phase I/II trial using a CXCR4 inhibitor in glioblastoma. Clin Cancer Res 2019;25(23):6948–57.
[115] Jordan JT, Sun W, Hussain SF, DeAngulo G, Prabhu SS, Heimberger AB. Preferential migration of regulatory T cells mediated by glioma-secreted chemokines can be blocked with chemotherapy. Cancer Immunol Immunother 2008;57(1):123–31.
[116] Zhu X, Fujita M, Snyder LA, Okada H. Systemic delivery of neutralizing antibody targeting CCL2 for glioma therapy. J Neuro-Oncol 2011;104(1):83–92.
[117] Finn RS, Qin S, Ikeda M, Galle PR, Ducreux M, Kim TY, et al. Atezolizumab plus Bevacizumab in Unresectable Hepatocellular Carcinoma. N Engl J Med 2020;382(20):1894–905.
[118] Kurz SC, Cabrera LP, Hastie D, Huang R, Unadkat P, Rinne M, et al. PD-1 inhibition has only limited clinical benefit in patients with recurrent high-grade glioma. Neurology 2018;91(14):e1355–9.
[119] Reardon DA, Nayak L, Peters KB, Clarke JL, Jordan JT, De Groot JF. Phase II study of pembrolizumab or pembrolizumab plus bevacizumab for recurrent glioblastoma patients. J Clin Oncol 2018;36(15 suppl):2006.

CHAPTER

2

Applied cancer immunogenomics in glioblastoma

Connor J. Liu[a,b], *Andrew T. Coxon*[b], *and Gavin P. Dunn*[a,b]

[a]Department of Neurological Surgery, Washington University School of Medicine, St. Louis, MO, United States [b]Andrew M. and Jane M. Bursky Center for Human Immunology and Immunotherapy Programs, Washington University School of Medicine, St. Louis, MO, United States

Abstract

The use of novel immunotherapeutic agents in treating malignant brain tumors faces considerable challenges. Cancer immunogenomics represents a genomic framework for studying the dynamic interactions between the host immune system and tumor microenvironment. Characterization of the tumor-specific antigen landscape has already yielded a multitude of clinical trials testing personalized glioblastoma vaccines. Furthermore, significant advancements in high-throughput sequencing technologies and bioinformatic analyses have enabled high-resolution analysis of local and peripheral adaptive immune repertoires in the setting of central nervous system malignancies. These robust methods offer exciting new techniques for quantifying antitumor immune responses that can be harnessed as biomarkers for response to immune treatment or as a means to design cell therapy strategies. In the future, immunogenomic studies will be critical to understanding the genomic basis of immunotherapy response and resistance and to inform the next generation of immunotherapy trials.

Abbreviations

CAR	chimeric antigen receptor
cDNA	complementary deoxyribonucleic acid
CDR	complementarity determining regions
CNS	central nervous system
DC	dendritic cell
dCTP	deoxycytidine triphosphate
DNA	deoxyribonucleic acid
EGFR	epidermal growth factor receptor
EGFRvIII	epidermal growth factor receptor variant III
ELISPOT	enzyme-linked immune absorbent plot
FDA	Food and Drug Administration

Immunotherapeutic Strategies for the Treatment of Glioma
https://doi.org/10.1016/B978-0-12-819755-4.00014-X

GAPVAC	Glioma Actively Personalized Vaccine Consortium
GBM	glioblastoma
GM-CSF	granulocyte-macrophage colony-stimulating factor
HLA	human leukocyte antigen
HTS	high-throughput sequencing
ICB	immune checkpoint blockade
IDH1	isocitrate dehydrogenase 1
ITGB1	integrin subunit beta 1
KLH	keyhole limpet hemocyanin
MGMT	O6-methylguanine-methyltransferase
MHC	major histocompatibility complex
MMR	mismatch repair
MS	mass spectrometry
NP-barcoded NACS	nanoparticle-barcoded nucleic acid cell sorting
NSCLC	non-small-cell lung cancer
PBMC	peripheral blood mononuclear cell
PCR	polymerase chain reaction
PD-1	programmed death 1
PSSM	position-specific scoring matrices
RNA	ribonucleic acid
TAP	transporter-associated with antigen processing
TCGA	The Cancer Genome Atlas
TCR	T cell receptor
TGF	transforming growth factor
TIL	tumor-infiltrating lymphocytes
TMG	tandem minigenes
TNF	tumor necrosis factor

Conflict of interest

No potential conflicts of interest were disclosed.

Introduction

In the past decade, immunotherapy has transformed the field of cancer treatment, with an explosion of clinical trials paving the way for FDA-approved immune-based cancer therapies and a Nobel prize in medicine awarded for the discovery of immune checkpoint-blocking agents. Immunotherapeutic strategies to target tumor antigens, such as cancer vaccines and adoptive transfer of tumor-infiltrating lymphocytes, have benefited immensely from recent advancements in high-throughput sequencing and associated bioinformatic analyses. However, as immunotherapy rapidly becomes the latest pillar of anticancer treatment, more work is needed to extend the efficacy of immune-based treatments for brain tumor patients. Indeed, there are currently no FDA-approved immune therapies for glioblastoma. Results from recent immunotherapy trials in glioblastoma raise questions about the immunological particularities of the central nervous system (CNS) [1–3]. While it is now appreciated that the CNS is not as hermetically immunoprivileged and immunologically inaccessible as previously thought [4–6], there are still major gaps in our understanding of fundamental components of the immune response to brain tumors, such as the identification of critical antigen-presenting cells, the location of naïve T cell priming, and the details underlying effector T cell trafficking into brain tumors, among others.

The central hypothesis underpinning immune-based treatment of cancer is that cancer cells are recognized as distinct from normal cells by the immune system. This immunologic discrimination may leverage the somatic mutations of the tumor. Specifically, tumors may present tumor-specific mutant antigens, termed neoantigens, bound to HLA molecules on the cell surface that can be recognized by tumor-specific T cells. Although much of the seminal work in describing neoantigens was performed in the late 20th century [7, 8], the field has been rapidly propelled forward by recent advancements in high-throughput sequencing and bioinformatic analysis of cancer genomes. These technological milestones coupled with a deeper understanding of the interplay between malignant progression and the immune system provide the basis of the modern field of cancer immunogenomics, which will be reviewed in this chapter. Cancer immunogenomic approaches prioritize identifying antigenic tumor mutations and leveraging components of the endogenous immune system that can be adapted for therapeutic purposes (Fig. 1) [5, 9]. Here, we will review the process of neoantigen identification as well as examine the role of neoantigens as predictive biomarkers and as immunotherapy targets. Finally, we will discuss contemporary methods to study T cell receptor repertoire dynamics in cancer.

Defining neoantigens

It has now been well studied that T cell recognition of tumor antigens presented on major histocompatibility complexes (MHC) represents an endogenous event in many cancer types [10–12]. These complex recognition mechanisms are mediated between highly specific interactions between T cell receptors and MHC-bound tumor antigens. To date, three general classes of tumor antigens have been documented: (1) shared tumor antigens, which are nonmutant, normally expressed proteins that are aberrantly overexpressed in tumor cells; (2) cancer testis antigens that are normally only found in healthy adult germ cell tissues but exhibit re-expression in some cancers; and (3) tumor-specific mutant antigens, referred to as neoantigens, which represent novel peptide sequences encoded by somatic mutations in the cancer genome. Although early clinical trials utilizing peptide-based vaccines comprising shared or cancer testis antigens have been unsuccessful, recent experimental efforts have leveraged the unique immunological properties of neoantigens in new cancer therapeutics [13–18]. As neoantigens are not expressed in normal tissues, T cells specific for those antigens should not be influenced by central immune tolerance. This specificity of neoantigens as immunological targets is critical to minimizing off-target effects of potential immune therapies. Furthermore, shard and cancer testis antigens have limited expression in tumor cells, unknown binding affinities with infrequent HLA alleles, and increased risk of "off-target" effects on nonmalignant cells. Due to their unique status as mutant antigens present only in the tumor, neoantigen-directed therapies should mitigate these autoimmune effects and

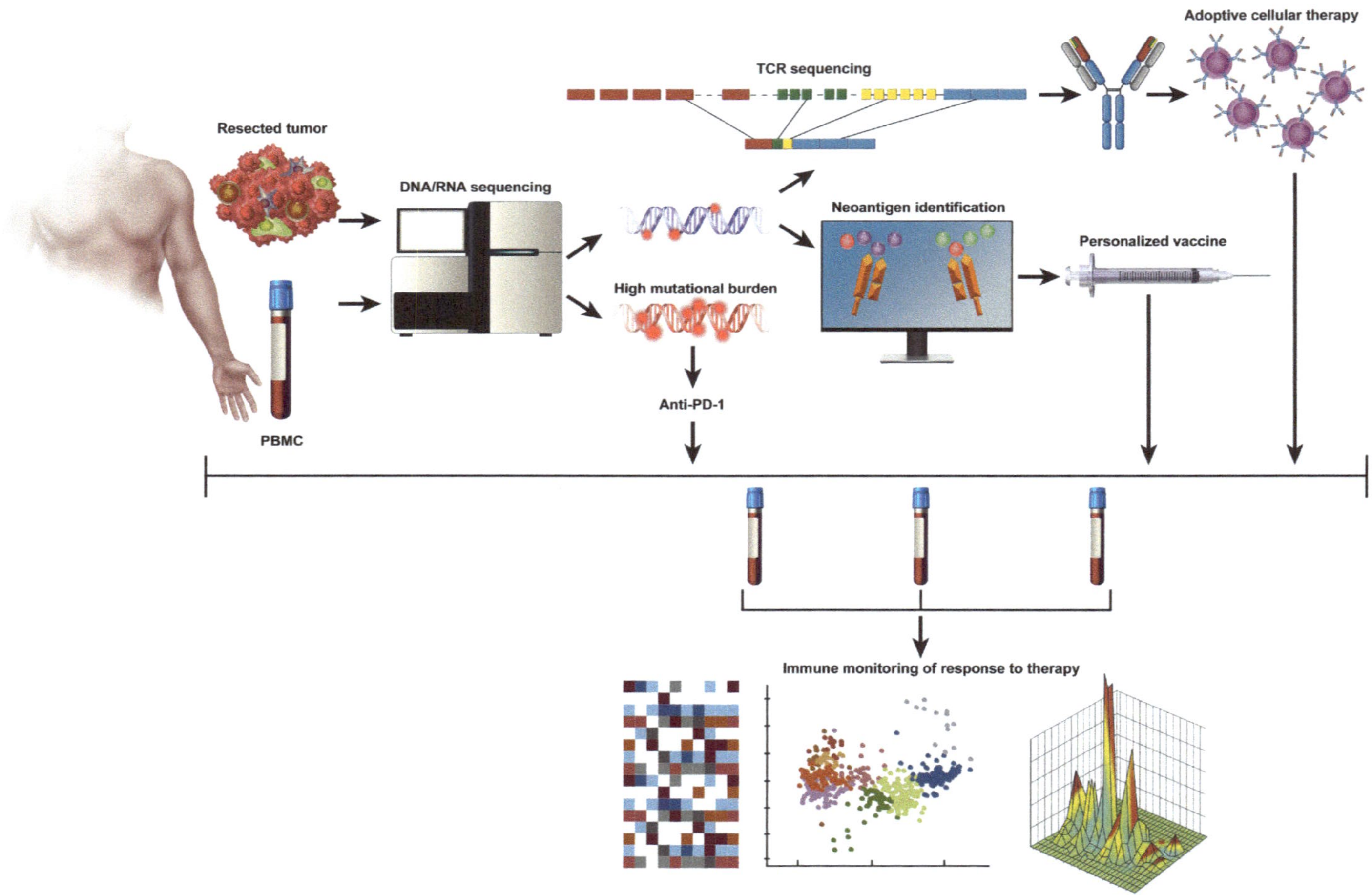

FIG. 1 Genomic analysis for personalized immunotherapeutics. Immunogenomic analysis begins with genomic sequencing to identify somatic tumor variants. RNA sequencing is employed to capture transcriptome expression data. Computational algorithms are used to generate personalized neoantigen predictions for therapeutic vaccination and for patient selection for checkpoint inhibitor therapy. Collection and genomic analysis of patient peripheral blood may also be used to profile T cell repertoires. In turn, T cell repertoire analysis provides a real-time data for monitoring patient responses to immunotherapy.

immune tolerance. Additionally, immunodominant neoantigens are tailored to a patient's specific HLA type, which allows the application of an identification pipeline to a larger population [19].

Discovering neoantigens: DNA/RNA sequencing

Whole-exome sequencing of tumor and matched normal tissue is essential to identify somatic tumor mutations [20, 21]. The tumor genome approach provides a comprehensive view of tumor mutations in the coding and noncoding regions. However, more cost-effective whole-exome sequencing approaches are commonly used and also aid in streamlining downstream sequence analysis [22]. Tumor RNA sequencing or cDNA-capture sequencing is then used to quantify transcript expression of mutant neoantigens. While early work in this area prioritized nonsynonymous missense mutations as neoantigen sources, recent developments in novel bioinformatic approaches enable the tumor genome to be mined for alternative aberrant events that may give rise to additional neoantigens [23]. These techniques allow researchers to evaluate the immunogenic potential of putative neoantigens derived from frameshift, splice site variants, gene fusion events, and endogenous retro-elements.

In silico approaches to neoantigen discovery

MHC class I predictions

Immunogenomic analysis of the tumor exome and transcriptome can generate thousands of potential mutant neoantigens. Of the thousands of candidates, only a handful may be immunogenic. To generate an effective immunotherapy, it is important to filter out the neoantigens that will not elicit T cell responses. In silico algorithms can predict peptide affinities to MHC alleles for individual patients [24]. Humans inherit 3 HLA class I (HLA-A, -B, and -C) and 3 HLA class II (HLA-DQ, -DP, and -DR) alleles from each parent. Combinations of HLA alleles allow 1 individual to express up to 12 different HLA complexes, which can each present its own set of neoantigens [9]. Furthermore, current HLA databases list over 6000 distinct alleles. Any pipeline to generate a neoantigen-based immunotherapy requires a prediction algorithm to cope with the vast genetic diversity of neoantigens and HLA allele combinations.

Most computational prediction algorithms for neoantigen binding use statistical models and machine learning algorithms. Early prediction algorithms calculate binding affinities using position-specific scoring matrices (PSSMs), which depend on experimental peptide-binding data for MHC alleles [25]. One of the available databases, the Immune Epitope Database, has over 120,000 epitopes and is widely used to discover new neoantigens [25, 26]. Despite the work that has gone into generating large databases for PSSM algorithms, they cannot optimally predict binding affinities for less common HLA alleles that do not have sufficient peptide:MHC class I binding data. New machine learning tools like NetMHC and its cross-species counterpart NetMHCpan use neural networks to predict novel peptide:MHC class I binding data from "training" on available data. NetMHCpan correctly identified 27/31 naturally occurring epitopes as immunogenic with a conventional binding affinity cutoff of $IC_{50} < 500\,nM$ [27]. While high peptide:MHC class I binding affinity is the most selective part of the antigen presentation pathway, some low binding affinity epitopes are still immunogenic [28]. Furthermore, the cutoff $IC_{50} < 500\,nM$ is controversial because a strong neoantigen may have a lower predicted affinity than observed due to rare HLA haplotypes skewing the neural net training [29].

MHC class II predictions

Accurate MHC class II in silico predictions are much harder to generate due to the nature of the proteins and the lack of validated neural net training sets. One barrier is that the general population expresses extensive HLA class II polymorphism [30]. Additionally, MHC class II molecules have an open peptide binding groove, as opposed to the closed groove of MHC class I molecules. The open ends of the class II groove allow for much longer peptides to bind, and these peptides can "slide" along the groove. While prediction algorithms only need to evaluate the peptide-binding region for MHC class I epitopes, class II prediction algorithms must also assess flanking amino acids outside of the biding groove [31, 32]. This complexity significantly decreases the accuracy of MHC class II neoantigen predictions.

Data showing the important contribution of $CD4^+$ T cells in the antitumor immune response reinforces the importance of generating better MHC class II predictions. T cell responses to a neoantigen vaccine in a mouse model of melanoma were 95% $CD4^+$ [33]. Furthermore, neoantigen-specific human $CD4^+$ T cells observed in a patient with melanoma and adoptive transfer of expanded neoantigen-specific $CD4^+$ T cells caused tumor regression and disease stabilization [34]. Most recently, Alspach et al. used a novel predictive algorithm incorporating both predicted binding affinities and mutation transcript expression levels to identify the MHC class II-restricted mutant ITGB1 neoantigen in the T3 sarcoma cell line [35]. They demonstrated the presence of mITGB1-specific $CD4^+$ T cells and ultimately show that prophylactic vaccination with both the class II and a previously characterized class I neoantigen was protective against T3 tumor challenge, whereas class I or II neoantigen vaccination alone was not. Together, their results support the role of $CD4^+$ T cells in priming robust $CD8^+$ antitumor responses and hold important implications for cancer vaccine strategies stimulating both $CD4^+$ and $CD8^+$ T cell responses.

Discovering neoantigens: Antigen processing

Prior to MHC presentation on the cell surface, antigens undergo a number of processing steps, including proteasomal processing and TAP protein transfer to the endoplasmic reticulum. A variety of computational tools have been designed to quantify these steps in order to calculate the probability a given antigen will reach the cell surface. These toolkits are continuously being updated and added to, and include, such programs as NetChop (proteasomal cleavage) and PredTAP and SVMTAP (TAP interactions) [34]. Using a NetChop filter with a peptide:MHC class I accurately identifies more neoantigens than NetMHC alone [36]. Applied together, these various computational prediction algorithms can help in focusing downstream neoantigen screening efforts on high-priority candidate antigens. However, as these additional neoantigen metrics have yet to be fully validated, researchers must exercise caution to not filter out antigens based on an individual score alone.

Discovering neoantigens: Mass spectrometry

Although the tumor transcriptome provides evidence of protein expression, mass spectrometry (MS) remains the most accurate albeit labor-intensive method to study antigen-MHC interactions. Whereas in silico predictions often have high sensitivity with low specificity, MS has high specificity but low sensitivity due to limitations in detection. The combination of high sensitivity in silico and high specificity in MS suggests that combining these approaches may yield more accurate neoantigen predictions. Tumor-bearing mice vaccinated against neoantigens predicted by both in silico algorithms and MS analysis lead to therapeutic T cell response [37]. When applied to a small cohort of patients with melanoma, MS identified 11 mutant peptides that were not among the top 10 predicted peptides by NetMHC [38]. Of the MS neoantigens, two elicited patient T cell responses. These data suggest that MS could help refine in silico prediction algorithms, especially for uncommon HLA haplotypes. Roadblocks to MS use clinically include the large amount of tissue required for analysis and the technical expertise necessary to operate MS machinery. Future advancements in MS equipment will hopefully alleviate these current obstructions.

Screening neoantigen reactivity

The in silico neoantigen discovery techniques described above may produce hundreds of potential neoantigen targets for a given tumor, each representing a computational prediction requiring further experimental validation. The goal of subsequent in vitro validation is to determine the immunogenicity of a given neoantigen and to establish whether an endogenous neoantigen-specific immune response exists. Enzyme-linked immune absorbent plot (ELISPOT) assays are a commonly employed method for evaluating functional neoantigen-specific T cell responses. Using ELISPOT assays, antigen-presenting cells flooded with a particular antigen are co-cultured with endogenous T cells, and immunogenicity is determined by the resultant T cell cytokine output such as interferon-γ, TNF-α, or TGF-β. More recently, novel techniques have been developed to address the laborious and time-intensive nature of screening large quantities of candidate antigens. The use of tandem minigenes (TMG) to screen patient neoantigens was first demonstrated by Lu and colleagues in a study of two melanoma patients who had experienced durable remissions following adoptive TIL therapy [39]. Using this approach, instead of testing individual antigens separately TMG DNA constructs encoding up to 24 neoantigens are transfected into autologous antigen-presenting cells and co-cultured with patient T cells. In the absence of a T cell response, the neoantigens contained in that TMG can be disregarded, while a positive signal would require further deconvolution of the pooled neoantigens to determine immunogenicity.

Peptide-MHC tetramers for multiplex flow cytometry may also be used to detect the presence of neoantigen-specific T cell receptors. Tetramers, which are comprised of four MHC molecules linked to a streptavidin-coated bead, can be designed for specific peptide sequences and used in a manner similar to fluorescent antibody staining for traditional flow cytometry. MHC tetramers for mass cytometry or DNA-labeled tetramers for sequencing analysis have also been described [40–42]. Compared with ELISPOT assays, tetramer-based techniques provide more granular T cell quantification and enable longitudinal tracking of neoantigen-specific T cell populations. However, one major drawback to these approaches is that they require large cell populations and are therefore typically used to analyze in vitro expanded T cells [43].

However, the identity of immunogenic neoantigens represents only one side of the immune recognition event, and given breakthroughs in adoptive cell therapies and CAR-T technology, there is an increasing need to profile neoantigen-specific T cell receptors. Peng and colleagues recently introduced their novel method using nanoparticle-barcoded nucleic acid cell sorting (NP-barcoded NACS) to identify neoantigen-specific T cell populations from nonexpanded tumor infiltrates or peripheral blood, and for pairing those neoantigen-specific CD8+ T cell populations with the cognate TCRα/β gene, using single-cell sequencing methods [43]. In this approach, peptide-MHC tetramers are coupled to magnetic nanoparticles via single-stranded DNA linkers. Nanoparticles loaded with up to 20,000 tetramers allow for the sensitive detection of neoantigen-specific T cells, and DNA barcodes enable pairing of antigens with their cognate TCR-α and TCR-β heterodimers.

Neoantigens as biomarkers

Characterizing the tumor neoantigen landscapes holds important implications for understanding mechanisms of immunotherapy response and resistance. Immune-modulating therapies such as checkpoint inhibition enhance immune recognition of tumor cells through an interaction mediated by tumor-specific neoantigens. Thus far, checkpoint inhibition therapy has proven most effective in tumors with moderate to high neoantigen burdens, leading to early FDA approval for their use in skin, lung, and bladder cancers [44]. The concept of high neoantigen burden and response to checkpoint blockade was further advanced by a study from Le and colleagues in colorectal cancer patients where they demonstrated robust responses to anti-PD-1 therapy in patients whose tumors were deficient in the DNA mismatch repair pathway compared with those with proficient machinery [45]. This study effectively established a correlation between high mutational burden, and by extension high neoantigen burden, with response to checkpoint inhibitor therapy (Fig. 1). While glioblastoma tumors typically contain a low to moderate numbers of mutations on the order of 70–100, a small minority of tumors exhibit a hypermutated genome containing over 1000 somatic mutations [46]. Over the years, two distinct pathways have been attributed to the hypermutant glioma phenotype: (1) a rare subset of newly diagnosed glioblastomas that contain germline or somatic mutations in either mismatch repair (MMR) or DNA polymerase genes, or (2) recurrent disease following temozolomide therapy. Consequently, there have been efforts to determine if hypermutant gliomas, which comprise a small percentage of primary tumors but approximately 20%–30% of posttreatment recurrent disease, may be a population that benefits from checkpoint inhibition. Anecdotally, there have been case reports of patients with hypermutated glioblastoma demonstrating dramatic responses to anti-PD-1 therapy after recurrence on standard of care [46, 47]. Several clinical trials are now underway to systematically address the question of checkpoint blockade efficacy in hypermutant glioma. While results from these trials are pending, data from a recent retrospective analysis of hypermutant recurrent glioma treated with pembrolizumab suggests that expectations for these trials should be limited. Touat and colleagues showed that in patients with hypermutated recurrent glioma, treatment with pembrolizumab was in fact less effective compared with other systemic therapies [48]. These unexpected findings warrant additional approaches to study glioma-immune interactions and suggest that answers will likely be multifactorial. Despite the abundance of predicted neoantigens present in hypermutated tumors, little is known about the immunogenicity of these putative targets. Further characterization of neoantigen-mediated immune responses in hypermutant disease is needed to assess the importance of neoantigen quality versus quantity. Additionally, the absence of a robust immune response to gliomas likely results from several complex aspects of the tumor microenvironment, and our understanding of these mechanisms will be key to enhancing antitumor immunity within the CNS.

Neoantigen as targets

Therapeutic targeting of neoantigens represents a promising new application of immunogenomics in cancer therapy. New efforts to explore neoantigen-directed therapies, such as cancer vaccines, are supported by substantial evidence that T cell-mediated recognition of tumor neoantigens is required for an effective antitumor immune response. Neoantigen-specific T cell responses have been shown to parallel patient tumor regression in response to anti-PD-1 treatment, suggesting that checkpoint blockade enhances neoantigen-specific T cell reactivity [43, 49]. These observations are supported by in vivo studies showing that neoantigen-reactive T cells can kill target-expressing tumor cells [50]. Furthermore, the loss of tumor neoantigens has been implicated as a mechanism of acquired resistance to immune checkpoint blockade. In the setting of malignant brain tumors, with the exception of a subset of hypermutant gliomas, most of these tumors harbor modest neoantigen burdens relative to those where checkpoint inhibition has proven most effective, such as NSCLC, melanoma, and bladder cancers. Thus while checkpoint blockade monotherapy may not represent a viable strategy for all brain tumors, there is optimism that neoantigen-based cancer vaccines, used alone or in combination with other immune-modulating agents, may yield greater success.

Epidermal growth factor receptor variant III

Epidermal growth factor receptor variant III (EGFRvIII) is a mutated EGFR present in 20%–30% of GBM. The EGFRvIII mutant receptor is constitutively active and may promote tumor growth through a paracrine mechanism [51]. EGFRvIII has an in-frame 801-base pair deletion in the extracellular domain. The new junctional region formed by the deletion forms a neoantigen. The rindopepimut vaccine targeted this junctional region with a 14-mer peptide spanning the region. Phase II clinical trials in patients with GBM using rindopepimut combined with standard-of-care temozolomide and radiation showed benefits in overall and progression-free survival compared with matched historical controls [52, 53]. However, the phase III trial of rindopepimut with temozolomide failed to show a survival benefit [54].

This trial did have several concerns that must be noted to guide future therapeutic vaccine efforts. First, EGFRvIII is only expressed in approximately 30% of GBM patients, and of the patients with the mutation, the EGFRvIII variant is heterogeneously expressed throughout the tumor [55]. Further, the early-phase studies showed that EGFRvIII-negative subclones emerged after vaccination. These data suggest that escape pressures from vaccinating against a single subclone may eventually render the immunotherapy ineffective. Further, it is important to note that the control group had better-than-expected survival compared with historical controls, which may suggest a small but significant benefit of immunostimulatory adjuvants like KLH and/or GM-CSF. It may be necessary to vaccinate against multiple targets, or pick targets that have high levels of expression throughout the tumor. Additionally, immune monitoring showed a predominant humoral response to rindopepimut. In addition to polyvalent vaccines, HLA binding predictions may enhance antitumor immunity through selectively stimulating cellular immunity. The lessons from rindopepimut have laid the foundation for future neoantigen vaccines.

Mutant IDH1 R132H

One potential neoantigen target is the mutant isocitrate dehydrogenase 1 (IDH1) gene. More than 70% of grade II, grade III, and secondary GBMs with mutant IDH1 have a mutation at R132 [56]. Of the IDH1 R132 mutants, about 90% are R132H. Furthermore, IDH1 mutations are thought to occur early in transformation, so IDH1 mutations should be relatively homogenous throughout the tumor cells. Mutant IDH1 is a tantalizing vaccination target since it is clonal and broadly expressed throughout the tumor. Immunizing mice against IDH1 R132H induces antitumor immunity, and some patients with the mutation have $CD4^+$ T cell responses against IDH1 R132H [57]. As expected, in silico analysis predicts that IDH1 R132H peptides have high affinity to MHC class II, consistent with the observed $CD4^+$ T cell response against IDH1 R132H. These data suggest that IDH1 R132H may be a good target for neoantigen vaccination and highlight the need to generate better MHC class II neoantigen binding predictions. Ongoing trials of IDH1 R132H mutant vaccination include a peptide vaccine administered with imiquimod in grade III–IV glioma (NCT02454634), a peptide vaccine administered with GM-CSF in grade II glioma (NCT02193347), and an autologous dendritic cell and cytotoxic lymphocyte-based vaccine in glioma (NCT02771301).

Personalized neoantigen vaccines

In addition to neoantigens common across multiple patients, the sequencing techniques and neoantigen prediction algorithms described above can be used to identify vaccine targets (Fig. 1) [58, 59]. Neoantigen vaccines have successfully elicited immune responses in patients with melanoma [17, 60]. These studies showed neoantigen-specific $CD4^+$ and $CD8^+$ T cell immune responses to peptide and RNA vaccines [61]. Multiple groups using neoantigen vaccines against glioblastoma have recently demonstrated safety, feasibility, and immunogenicity.

Johanns and colleagues reported on a single patient treated with a personalized vaccine made of autologous tumor lysate pulsed dendritic cells followed by the administration of eight synthetic long peptides corresponding to tumor neoantigens with poly-ICLC adjuvant [15]. Additionally, the "Glioma Actively Personalized Vaccine Consortium" (GAPVAC) study treated 15 glioblastoma patients with personalized dual-product vaccines, radiation, and temozolomide [62]. The first vaccine (APVAC1) included up to seven best ranked "off-the-shelf" tumor-associated antigen peptides. The antigens were ranked based on tumor expression, HLA binding, and immunogenicity. The second vaccine (GAPVAC2) included two neoantigen peptides identified by sequencing and HLA-binding affinity. Lastly, Keskin and colleagues treated eight MGMT unmethylated glioblastoma patients with 7–20 neoantigen peptides identified through sequencing and HLA binding prediction [16]. The patients were vaccinated on a prime/boost schedule following radiotherapy, but temozolomide was held due to the marginal benefit in MGMT unmethylated glioblastoma. All three studies demonstrated neoantigen-specific T cells following vaccination in peripheral blood. Tumor tissue resected after recurrence also showed neoantigen-specific T cell infiltrate in each study. Of note, Keskin and colleagues reported that patients on concomitant dexamethasone did not have detectable responses to immunized neoantigen peptides, while patients not on dexamethasone generated tumor-specific $CD4^+$ and $CD8^+$ T cell responses to multiple neoantigen peptides.

Overall, these studies support the feasibility and safety of personalized neoantigen vaccines. These studies also confirm that personalized neoantigen vaccines generate neoantigen-specific T cell responses that infiltrate the tumor. Additionally, dexamethasone administration likely attenuates vaccine responses and should be used only when necessary. Patients who will likely require corticosteroids or other immunosuppression may not be good candidates for these therapies. Although trials with checkpoint blockade monotherapy in glioblastoma have yielded mixed results, multimodal strategies aimed at driving effector immune cells into the tumor coupled with strategies to alleviate local immune suppression represent promising new approaches. Indeed, Liu and colleagues recently demonstrated that in a checkpoint blockade-resistant murine glioblastoma, a combination of polyvalent neoantigen vaccination and checkpoint inhibition provided a synergistic survival benefit compared with either treatment alone [63]. Ongoing studies evaluating neoantigen vaccination for glioblastoma patients are combining vaccination with PD-1 checkpoint inhibition (NCT02287428) and evaluating a novel DNA-based plasmid vector (NCT04015700). Future studies will illuminate the functionality of the neoantigen-specific tumor-infiltrating lymphocytes. These and future studies will assess the platforms and combination therapies necessary to generate the optimal antitumor response.

T cell receptor (TCR) repertoire profiling

As the field of immuno-oncology evolves, novel therapeutic strategies aimed at harnessing antitumor T cell cytotoxicity have shown promising results, from personalized neoantigen vaccines and oncolytic viruses, to immune checkpoint blockade (ICB) and adoptive cell transfer (ACT) therapy. Given the central role of T cells in cancer immunotherapy, a deeper understanding of T cell genetic programs, antigen specificity, and repertoire diversity is critical to new therapeutic development and immune treatment monitoring. Recent advances in high-throughput sequencing (HTS) technologies have facilitated experimental efforts to examine T cell receptor repertoires at a single-nucleotide resolution and at a scale large enough to observe population-wide changes.

As one of the principal sites for antigen recognition, T cell receptors (TCR) exhibit immense diversity needed to recognize and neutralize a vast array of extrinsic threats. During T cell development, variable antigen receptor gene segments (variable (V), joining (J), diversity (D)) are rearranged through targeted DNA recombination events [62]. Additional pseudo-random editing is introduced by the addition or deletion of nucleotides at the junctions of these segments. While the entire variable region shapes receptor function, sequences within the complementarity determining regions (CDRs), and CDR3 in particular, contribute most to TCR specificities. As this recombination process occurs separately for both α and β subunit chains, subsequent heterodimeric pairing brings forth still greater combinatorial diversity. Taken together, the diversity created through these programmed mechanisms is vast, with the theoretical number of distinct αβ TCRs estimated to exceed 10^{18} [64]Thus, the TCR nucleotide sequence provides a nearly unique molecular tag and offers functional information regarding antigen specificity for each clone in the adaptive immune system.

TCR sequencing approaches

In order to capture the vast diversity present in a typical T cell repertoire, TCR profiling studies may employ analyses at either the bulk TCR or single-cell sequence level [64]. While single-cell TCR analyses offer the advantage of accurately identifying paired α and β subunit chains, bulk sequencing approaches are currently more widely employed due to lower cost and ease of obtaining adequate samples from clinical specimens (whereas single-cell sequencing requires fresh material for sorting of live cells), without compromising the ability to quantify repertoire diversity. However, even as HTS technology becomes increasingly accessible and routine for genomic and transcriptomic sequencing, its application to study lymphocyte repertoires poses unique technical challenges due to the extensive diversity it aims to capture. Furthermore, the bioinformatic workflows required to interpret raw sequence data continue to evolve. Therefore, there is no single approach to immune repertoire profiling that is appropriate for all studies, and recommendations regarding genomic library construction, extent of receptor coverage necessary, minimization of amplification bias, as well as data analysis methods, continue to develop.

Current immune repertoire profiling methodology can be generally divided into three separate stages: (1) library preparation and sequencing; (2) lower-level processing such as sequence assembly, assignment to genomic V, D and J genes, and extraction of CDR3 regions; and (3) higher-level processing and analysis. Library preparation consists of amplifying recombined TCR genes from either genomic DNA or cDNA using multiplex polymerase chain reaction (PCR). Combining constant region primers and a mix of primers for all known V alleles allows for specific amplification of the TCR transcript across the CDR3 region [65]. While both genomic DNA and mRNA may be used as input for this initial step, the choice of raw genetic materials may be tailored to the needs of individual experiments. Starting with genomic DNA has a number of advantages: (a) Samples are typically easier to obtain as biopsy samples from tissues or slides can be used and (b) the stable copy number of genomic DNA in each cell allows for more reliable quantification of lymphocyte clonal expansions within diverse repertoires. In contrast, when using mRNA, quantification of single TCR clones is more challenging as a cell will contain multiple TCR transcripts. However, many current methods are designed for RNA, as RNA TCR products do not contain introns and therefore allow for more complete capture of CDR3 regions. RNA transcripts are also more abundant than their template genomic DNA, allowing for the sensitive detection of less-represented sequences [66].

Amplification biases that occur during multiplex PCR can skew the perceived frequency of particular alleles and distort the relative transcript abundance [67]. It is possible to correct for this kind of error by using a specific experimental design, including adjustment of primer concentrations and using molecular barcoding. For protocols using RNA samples, a rapid amplification of 5′ complementary DNA ends (5′RACE), employing the template-switch effect, is becoming a gold standard for bulk TCR analysis [62]. In this method, RNA is reverse-transcribed by an enzyme endowed with terminal transferase activity that adds a stretch of nontemplate dCTPs at the 3′ end of the cDNA [30]. A template-switch oligonucleotide containing an oligo(rG) sequence anchors to the nontemplate stretch of the first-strand cDNA, allowing the reverse transcriptase to switch templates and continue replicating to the end of the oligonucleotide. This enables the synthesis of cDNA strands containing the complete 5′ end of the mRNA, independent of the carried V allele, which enables unbiased capture of all TCR variants present in the sample.

Once TCR genes have been enriched, amplified, and sequenced, initial processing of sequence data involves sequence alignment; assignment to genomic V, D, and J genes; and extraction of CDR3 regions. The International Immunogenetics Information System, or IMGT, was first developed in 1989 as a centralized repository for immunogenetics-related data (www.imgt.org) [68]. It remains the most widely used reference point for immunogenetics, and specifically for maintaining the IMGT/GENE-DB database, which contains a well-annotated set of genomic V, D, and J genes from several species. This set of genes provides the reference for most TCR gene assignment tools. In addition to providing the reference sequences for TCR genes, the IMGT resource page also provides various tools for sequence assignment, including IMGT/V-QUEST, and a higher-throughput version, IMGT/HighV-QUEST [69]. Gene assignment uses global pairwise alignment algorithms to identify the best-fitting V, D, and J genes, and then the Smith-Waterman local alignment algorithm to determine the deletions at the end of the V(D)J genes, and the insertions between them.

Once raw sequence data has been processed, higher-level analyses can be performed to provide biological insight into the underlying immunology. High-level processing includes diversity measurements, clonal frequency distributions, analysis of differential V and J usage, and analysis of interindividual sharing of TCR sequences (public versus private). Despite the enormous number of unique sequences captured in a given experiment, essential information regarding TCR diversity and overlap between repertoires can be captured using just a small number of these aforementioned metrics. Furthermore, there exist a number of publicly available toolsets that have been validated for the purpose of performing specialized analyses and generating summary statistics. VDJtools is a software package for basic statistical analysis of immune repertoires and includes a number of other specialized programs, including MIXCR for fast alignment and clonal-type assembly, MIGEC for removing duplicates and combining barcodes, and VDJviz for visualization [70].

As described above, immune repertoire sequencing requires specific library preparation techniques that can be expensive with regard to financial costs as well as labor. To address these challenges, new computational approaches have been designed to study the repertoire of T and B cells using unselected RNA sequencing data. These new techniques enable researchers to mine the terabytes of sequencing data that has already been made publicly available on databases such as The Cancer Genome Atlas (TCGA). TRUST is a computational method for de novo assembly of CDR3 regions using paired-end RNA-seq data [71, 72]. TRUST analysis of 9142 samples from TCGA allowed researchers to characterize trends between mutational burden and T cell diversity, as well as to identify three potential immunogenic somatic mutations based on their co-occurrence with CDR3 sequences. Given the decreasing sequencing costs and increasing efforts to study tumor-immune profiles, computational techniques that facilitate large-scale genomic analyses will be critical to generating important biological insights that drive the field forward.

TCR repertoires in cancer

Advancements in high-throughput TCR sequencing technologies offer new ways to study complex interactions between cancer, the immune system, and treatment. In particular, understanding which brain tumor patients stand to benefit from immune-modulating therapies remains a critical task. Across cancer types, the detection of intratumoral clonal T cell expansion has been used as a surrogate for tumor-specific T cell responses [73]. However, in the context of CNS malignancies where there is a significant risk associated with obtaining tissue samples for analysis, less-invasive methods of monitoring TCR repertoires and response to treatment are necessary. Sims and colleagues first reported on the significant divergence of TCR repertoires, as measured by their CDR3 regions, between peripheral blood and intratumoral populations in patients with newly diagnosed glioma [74]. They used the degree of PBMC-TIL TCR divergence to assess antitumor immune responses, with low levels of divergence concomitant with repertoire narrowing typically observed with antigen-specific T cell responses. Significantly, they identified a signature set of CDR3s in the peripheral blood that could reliably predict the degree of PBMC-TIL divergence, representing a noninvasive method of monitoring intratumoral immune status (Fig. 1). In a similar pursuit, Hsu and colleagues used TCR repertoire profiling to monitor patient responses to a novel dendritic cell vaccine [75]. Glioblastoma patients enrolled in the trial were treated with an autologous tumor lysate-pulsed dendritic cell (DC) vaccine and underwent TCR repertoire profiling before and after treatment. They found that a greater degree of TCR sharing between peripheral blood and tumor correlated with improved patient survival. That this correlation held true both before and after DC vaccination highlights the utility of TCR sequencing as an exciting predictive tool.

In addition to its facility to predict and monitor responses to treatment, TCR profiling is being used to inform new therapeutic strategies using existing immune checkpoint blocking (ICB) agents. Zhao and colleagues carried out a retrospective analysis of patients with recurrent GBM treated with anti-PD-1 and reported on the evolution of intratumoral lymphocytes by tracking TCR clones [76]. Compared with patients with stable disease 6 months after treatment, nonresponders were found to have greater increases in T cell diversity after anti-PD-1 treatment, indicating impaired T cell recruitment into the tumor microenvironment. While overall anti-PD-1 therapy in this study did not provide a survival benefit, these findings support that a subgroup of patients may benefit from checkpoint inhibition and highlights a way forward for more personalized approaches to improve patient selection for immunotherapy.

In 2019, Schalper et al. and Cloughesy et al. published a pair of studies reporting on prospective clinical trials in which recurrent GBM was treated with neoadjuvant anti-PD-1. Cloughesy et al. found that in a randomized trial, patients treated with neoadjuvant anti-PD-1 (before and after surgery) had a greater overall and progression-free survival compared with patients who received adjuvant (after surgery) immunotherapy [16]. Interestingly, in the neoadjuvant group, there was a much greater degree of TCR clonotype overlap between tumor and blood, an observation that has also recently been found in recurrent GBM patients treated with neoantigen vaccines [77]. Schalper and colleagues set out to study how neoadjuvant anti-PD-1 modifies the immune milieu of glioblastoma compared with a historical control series of GBM patients who did not receive immunotherapy [78]. Although no significant effects on the composition of GBM microenvironment immune cells were seen with immunotherapy, TCR repertoire analysis showed more diversity in patients treated with neoadjuvant anti-PD-1 with an association between TCR diversity and survival. Together, all three studies suggest that neoadjuvant immune therapy results in local immune activation with specific effects on TCR repertoires. Furthermore, as there is still a lack of clarity regarding the effects of neoadjuvant checkpoint inhibition on intratumoral TCR diversity, larger studies will be critical to understanding the longitudinal evolution of GBM-specific T cell clones.

TCR profiling technologies are also being used to advance our understanding of immune correlates in disease progression. Disease recurrence remains inevitable for glioblastoma patients and comes with few effective therapies. Mohmes et al. utilized TCR sequencing to compare matched newly diagnosed and recurrent glioblastomas, and found that recurrent tumors occur with significantly fewer TCR clonotypes [78]. Furthermore, they found that less than 20% of TCR clones were shared between primary and recurrent samples, suggesting that clonal restriction at the time of disease progression was not due to a narrowing of preexisting clonotypes and may instead indicate T cell responses to new tumor antigens. The apparent reduction in available T cell tumor antigens during disease progression represents a considerable challenge for the checkpoint inhibitors trials in recurrent glioblastoma. These findings highlight the importance of studying tissue samples at multiple time points during disease and treatment. Within this framework, neoantigen-directed therapies, designed to specifically target new tumor subclones, could be manufactured and administered iteratively in order to sequester disease progression. Similar approaches are needed to understand the effects of standard treatments on immune responses to brain tumors. In the same study, the authors found that within peripheral blood samples taken at the time of disease diagnosis and progression, the degree of TCR repertoire overlap increased with patient age. The study authors hypothesize that the immune systems of older patients are less equipped to tolerate the effects of chemotherapy and radiation, further reducing their ability to reconstitute peripheral T cell diversity.

Conclusions and future perspectives

We have summarized promising new developments in cancer immunogenomics germane to GBM immunotherapy, including patient-specific neoantigen vaccination, TCR repertoire analysis for immune treatment monitoring, neoadjuvant checkpoint blockade immunotherapy, and CAR T cell therapies (Fig. 1). The ongoing work in these areas suggests that the future of immunotherapy for GBM is most hopeful for combination therapies. Equally as important will be the development of TCR profiling analyses and other methods of immune treatment monitoring to inform rational clinical trial design and prudent patient stratification. As high-throughput sequencing technologies advance and our understanding of immunology grows, the field of cancer immunogenomics will be pivotal to bringing more effective treatments to patients. These areas of research will remain important in order to understand the genomic basis of response and resistance, as well as to enable the targeting of tumor-specific neoantigens. While immunotherapy faces considerable challenges in GBM treatment and early clinical trials have been met with mixed results, emerging data in the field are generating new hope and are already being used to inform the next succession of immunotherapy trials. Indeed, there is a general sense that the field is still at the beginning, albeit unsure of where the end might be. Ultimately, as we expand on this work in larger studies and with advanced scientific approaches, we will learn not just about their therapeutic efficacy but also gain new insights into the fundamental immunobiology of the CNS itself.

Disclosures

G.P.D. is a co-founder of Immunovalent, which has no bearing on this manuscript.

Funding support

This work was supported by R01NS112712-01 (NINDS), K08NS092912-05 (NINDS), and Damon Runyon Cancer Research Foundation (G.P.D.).

References

[1] Liau LM, Ashkan K, Tran DD, et al. First results on survival from a large phase 3 clinical trial of an autologous dendritic cell vaccine in newly diagnosed glioblastoma. J Transl Med 2018;16(1):142. https://doi.org/10.1186/s12967-018-1507-6.

[2] Reardon DA, Omuro A, Brandes AA, et al. OS10.3 randomized phase 3 study evaluating the efficacy and safety of nivolumab vs bevacizumab in patients with recurrent glioblastoma: CheckMate 143. Neuro-Oncology 2017;19(suppl_3):iii21. https://doi.org/10.1093/neuonc/nox036.071.

[3] Omuro A, Vlahovic G, Lim M, et al. Nivolumab with or without ipilimumab in patients with recurrent glioblastoma: results from exploratory phase I cohorts of CheckMate 143. Neuro-Oncology 2018;20(5):674–86. https://doi.org/10.1093/neuonc/nox208.

[4] Dunn GP, Fecci PE, Curry WT. Cancer immunoediting in malignant glioma. Neurosurgery 2012;71(2):201–23. https://doi.org/10.1227/NEU.0b013e31824f840d.

[5] Johanns TM, Dunn GP. Applied cancer immunogenomics: leveraging neoantigen discovery in glioblastoma. Cancer J 2017;23(2):125–30. https://doi.org/10.1097/PPO.0000000000000247.

[6] Louveau A, Smirnov I, Keyes TJ, et al. Structural and functional features of central nervous system lymphatic vessels. Nature 2015;523(7560):337–41. https://doi.org/10.1038/nature14432.

[7] van der Bruggen P, Traversari C, Chomez P, et al. A gene encoding an antigen recognized by cytolytic T lymphocytes on a human melanoma. Science 1991;254(5038):1643–7. https://doi.org/10.1126/science.1840703.
[8] Sahin U, Türeci Ö, Schmitt H, et al. Human neoplasms elicit multiple specific immune responses in the autologous host. Proc Natl Acad Sci U S A 1995;92(25):11810–3. https://doi.org/10.1073/pnas.92.25.11810.
[9] Johanns TM, Bowman-Kirigin JA, Liu C, Dunn GP. Targeting neoantigens in glioblastoma: an overview of cancer immunogenomics and translational implications. Neurosurgery 2017;64(CN_suppl_1):165–76. https://doi.org/10.1093/neuros/nyx321.
[10] Kalaora S, Wolf Y, Feferman T, et al. Combined analysis of antigen presentation and T-cell recognition reveals restricted immune responses in melanoma. Cancer Discov 2018;8(11):1366–75. https://doi.org/10.1158/2159-8290.CD-17-1418.
[11] Ren Y, Cherukuri Y, Wickland DP, et al. HLA class-I and class-II restricted neoantigen loads predict overall survival in breast cancer. OncoImmunology 2020;9(1). https://doi.org/10.1080/2162402X.2020.1744947.
[12] Liu S, Matsuzaki J, Wei L, et al. Efficient identification of neoantigen-specific T-cell responses in advanced human ovarian cancer. J Immunother Cancer 2019;7(1). https://doi.org/10.1186/s40425-019-0629-6.
[13] Wong KK, Li WWA, Mooney DJ, Dranoff G. Advances in therapeutic cancer vaccines. 1st ed. vol. 130. Elsevier Inc.; 2016. https://doi.org/10.1016/bs.ai.2015.12.001.
[14] Hilf N, Kuttruff-Coqui S, Frenzel K, et al. Actively personalized vaccination trial for newly diagnosed glioblastoma. Nature 2019;565(7738):240–5. https://doi.org/10.1038/s41586-018-0810-y.
[15] Johanns TM, Miller CA, Liu CJ, et al. Detection of neoantigen-specific T cells following a personalized vaccine in a patient with glioblastoma. OncoImmunology 2019;8(4). https://doi.org/10.1080/2162402X.2018.1561106, e1561106.
[16] Keskin DB, Anandappa AJ, Sun J, et al. Neoantigen vaccine generates intratumoral T cell responses in phase Ib glioblastoma trial. Nature 2019;565(7738):234–9. https://doi.org/10.1038/s41586-018-0792-9.
[17] Ott PA, Hu Z, Keskin DB, et al. An immunogenic personal neoantigen vaccine for patients with melanoma. Nature 2017;547(7662):217–21. https://doi.org/10.1038/nature22991.
[18] Zolkind P, Dunn GP, Lin T, Griffith M, Griffith OL, Uppaluri R. Neoantigens in immunotherapy and personalized vaccines: implications for head and neck squamous cell carcinoma. Oral Oncol 2017;71:169–76. https://doi.org/10.1016/j.oraloncology.2016.09.010.
[19] Schumacher TN, Schreiber RD. Neoantigens in cancer immunotherapy. Science 2015;348(6230):69–74. https://doi.org/10.1126/science.aaa4971.
[20] Lee W, Jiang Z, Liu J, et al. The mutation spectrum revealed by paired genome sequences from a lung cancer patient. Nature 2010;465(7297):473–7. https://doi.org/10.1038/nature09004.
[21] Ley TJ, Mardis ER, Ding L, et al. DNA sequencing of a cytogenetically normal acute myeloid leukaemia genome. Nature 2008;456(7218):66–72. https://doi.org/10.1038/nature07485.
[22] Matsushita H, Vesely MD, Koboldt DC, et al. Cancer exome analysis reveals a T-cell-dependent mechanism of cancer immunoediting. Nature 2012;482(7385):400–4. https://doi.org/10.1038/nature10755.
[23] Zhang J, White NM, Schmidt HK, et al. INTEGRATE: gene fusion discovery using whole genome and transcriptome data. Genome Res 2016;26(1):108–18. https://doi.org/10.1101/gr.186114.114.
[24] Lin HH, Ray S, Tongchusak S, Reinherz EL, Brusic V. Evaluation of MHC class I peptide binding prediction servers: applications for vaccine research. BMC Immunol 2008;9(1):8. https://doi.org/10.1186/1471-2172-9-8.
[25] Pabinger S, Dander A, Fischer M, et al. A survey of tools for variant analysis of next-generation genome sequencing data. Brief Bioinform 2014. https://doi.org/10.1093/bib/bbs086.
[26] Yang X, Yu X. An introduction to epitope prediction methods and software. Rev Med Virol 2009;19(2):77–96. https://doi.org/10.1002/rmv.602.
[27] Fritsch EF, Rajasagi M, Ott PA, Brusic V, Hacohen N, Wu CJ. HLA-binding properties of tumor neoepitopes in humans. Cancer Immunol Res 2014;2(6):522–9. https://doi.org/10.1158/2326-6066.CIR-13-0227.
[28] Duan F, Duitama J, Al Seesi S, et al. Genomic and bioinformatic profiling of mutational neoepitopes reveals new rules to predict anticancer immunogenicity. J Exp Med 2014;211(11):2231–48. https://doi.org/10.1084/jem.20141308.
[29] Liu XS, Mardis ER. Applications of immunogenomics to cancer. Cell 2017;168(4):600–12. https://doi.org/10.1016/j.cell.2017.01.014.
[30] Robinson J, Waller MJ, Parham P, et al. IMGT/HLA and IMGT/MHC: sequence databases for the study of the major histocompatibility complex. Nucleic Acids Res 2003. https://doi.org/10.1093/nar/GKG070.

[31] Gubin MM, Artyomov MN, Mardis ER, Schreiber RD. Tumor neoantigens: building a framework for personalized cancer immunotherapy. J Clin Investig 2015;125(9):3413–21. https://doi.org/10.1172/JCI80008.

[32] Linnemann C, van Buuren MM, Bies L, et al. High-throughput epitope discovery reveals frequent recognition of neo-antigens by CD4+ T cells in human melanoma. Nat Med 2015;21(1):81–5. https://doi.org/10.1038/nm.3773.

[33] Kreiter S, Vormehr M, van de Roemer N, et al. Mutant MHC class II epitopes drive therapeutic immune responses to cancer. Nature 2015;520(7549):692–6. https://doi.org/10.1038/nature14426.

[34] Lu YC, Robbins PF. Cancer immunotherapy targeting neoantigens. Semin Immunol 2016;28(1):22–7. https://doi.org/10.1016/j.smim.2015.11.002.

[35] Alspach E, Lussier DM, Miceli AP, et al. MHC-II neoantigens shape tumour immunity and response to immunotherapy. Nature 2019;574(7780):696–701. https://doi.org/10.1038/s41586-019-1671-8.

[36] van Buuren MM, Calis JJA, Schumacher TNM. High sensitivity of cancer exome-based CD8 T cell neo-antigen identification. OncoImmunology 2014;3(5). https://doi.org/10.4161/onci.28836.

[37] Yadav M, Jhunjhunwala S, Phung QT, et al. Predicting immunogenic tumour mutations by combining mass spectrometry and exome sequencing. Nature 2014;515(7528):572–6. https://doi.org/10.1038/nature14001.

[38] Bassani-Sternberg M, Bräunlein E, Klar R, et al. Direct identification of clinically relevant neoepitopes presented on native human melanoma tissue by mass spectrometry. Nat Commun 2016;7(1):1–16. https://doi.org/10.1038/ncomms13404.

[39] Lu YC, Yao X, Crystal JS, et al. Efficient identification of mutated cancer antigens recognized by T cells associated with durable tumor regressions. Clin Cancer Res 2014;20(13):3401–10. https://doi.org/10.1158/1078-0432.CCR-14-0433.

[40] Bentzen AK, Marquard AM, Lyngaa R, et al. Large-scale detection of antigen-specific T cells using peptide-MHC-I multimers labeled with DNA barcodes. Nat Biotechnol 2016;34(10):1037–45. https://doi.org/10.1038/nbt.3662.

[41] Newell EW, Sigal N, Nair N, Kidd BA, Greenberg HB, Davis MM. Combinatorial tetramer staining and mass cytometry analysis facilitate T-cell epitope mapping and characterization. Nat Biotechnol 2013;31(7):623–9. https://doi.org/10.1038/nbt.2593.

[42] Fehlings M, Simoni Y, Penny HL, et al. Checkpoint blockade immunotherapy reshapes the high-dimensional phenotypic heterogeneity of murine intratumoural neoantigen-specific CD8+ T cells. Nat Commun 2017;8(1):1–12. https://doi.org/10.1038/s41467-017-00627-z.

[43] McGranahan N, Furness AJS, Rosenthal R, et al. Clonal neoantigens elicit T cell immunoreactivity and sensitivity to immune checkpoint blockade. Science (New York, NY) 2016;351(6280):1463–9. https://doi.org/10.1126/science.aaf1490.

[44] Ribas A, Wolchok JD. Cancer immunotherapy using checkpoint blockade. Science 2018;359(6382):1350–5. https://doi.org/10.1126/science.aar4060.

[45] Le DT, Uram JN, Wang H, et al. PD-1 blockade in tumors with mismatch-repair deficiency. N Engl J Med 2015;372(26):2509–20. https://doi.org/10.1056/NEJMoa1500596.

[46] Johanns TM, Miller CA, Dorward IG, et al. Immunogenomics of hypermutated glioblastoma: a patient with germline POLE deficiency treated with checkpoint blockade immunotherapy. Cancer Immunol Res 2016;4(11):1007–15. https://doi.org/10.1158/2326-6066.CIR-16-0156.

[47] Bouffet E, Larouche V, Campbell BB, et al. Immune checkpoint inhibition for hypermutant glioblastoma multiforme resulting from germline biallelic mismatch repair deficiency. J Clin Oncol 2016;34(19):2206–11. https://doi.org/10.1200/JCO.2016.66.6552.

[48] Touat M, Li YY, Boynton AN, et al. Mechanisms and therapeutic implications of hypermutation in gliomas. Nature 2020;580(7804):517–23. https://doi.org/10.1038/s41586-020-2209-9.

[49] Rizvi NA, Hellmann MD, Snyder A, et al. Mutational landscape determines sensitivity to PD-1 blockade in non-small cell lung cancer. Science 2015;348(6230):124–8. https://doi.org/10.1126/science.aaa1348.

[50] Castle JC, Kreiter S, Diekmann J, et al. Exploiting the mutanome for tumor vaccination. Cancer Res 2012;72(5):1081–91. https://doi.org/10.1158/0008-5472.CAN-11-3722.

[51] Inda MDM, Bonavia R, Mukasa A, et al. Tumor heterogeneity is an active process maintained by a mutant EGFR-induced cytokine circuit in glioblastoma. Genes Dev 2010;24(16):1731–45. https://doi.org/10.1101/gad.1890510.

[52] Swartz AM, Li QJ, Sampson JH. Rindopepimut: a promising immunotherapeutic for the treatment of glioblastoma multiforme. Immunotherapy 2014;6(6):679–90. https://doi.org/10.2217/imt.14.21.

[53] Sampson JH, Heimberger AB, Archer GE, et al. Immunologic escape after prolonged progression-free survival with epidermal growth factor receptor variant III peptide vaccination in patients with newly diagnosed glioblastoma. J Clin Oncol 2010;28(31):4722–9. https://doi.org/10.1200/JCO.2010.28.6963.

[54] Weller M, Butowski N, Tran DD, et al. Rindopepimut with temozolomide for patients with newly diagnosed, EGFRvIII-expressing glioblastoma (ACT IV): a randomised, double-blind, international phase 3 trial. Lancet Oncol 2017;18(10):1373–85. https://doi.org/10.1016/S1470-2045(17)30517-X.

[55] Nishikawa R, Sugiyama T, Narita Y, Furnari F, Cavenee WK, Matsutani M. Immunohistochemical analysis of the mutant epidermal growth factor, ΔEGFR, in glioblastoma. Brain Tumor Pathol 2004;21(2):53–6. https://doi.org/10.1007/BF02484510.

[56] Yan H, Parsons DW, Jin G, et al. IDH1 and IDH2 mutations in gliomas. N Engl J Med 2009;360(8):765–73. https://doi.org/10.1056/NEJMoa0808710.

[57] Schumacher T, Bunse L, Pusch S, et al. A vaccine targeting mutant IDH1 induces antitumour immunity. Nature 2014;512(7514):324–7. https://doi.org/10.1038/nature13387.

[58] Sahin U, Türeci Ö. Personalized vaccines for cancer immunotherapy. Science 2018;359(6382):1355–60. https://doi.org/10.1126/science.aar7112.

[59] Hu Z, Ott PA, Wu CJ. Towards personalized, tumour-specific, therapeutic vaccines for cancer. Nat Rev Immunol 2018;18(3):168–82. https://doi.org/10.1038/nri.2017.131.

[60] Carreno BM, Magrini V, Becker-Hapak M, et al. A dendritic cell vaccine increases the breadth and diversity of melanoma neoantigen-specific T cells. Science 2015;348(6236):803–8. https://doi.org/10.1126/science.aaa3828.

[61] Sahin U, Derhovanessian E, Miller M, et al. Personalized RNA mutanome vaccines mobilize poly-specific therapeutic immunity against cancer. Nature 2017;547(7662):222–6. https://doi.org/10.1038/nature23003.

[62] Rosati E, Dowds CM, Liaskou E, Henriksen EKK, Karlsen TH, Franke A. Overview of methodologies for T-cell receptor repertoire analysis. BMC Biotechnol 2017;17(1). https://doi.org/10.1186/s12896-017-0379-9.

[63] Liu CJ, Schaettler M, Blaha DT, et al. Treatment of an aggressive orthotopic murine glioblastoma model with combination checkpoint blockade and a multivalent neoantigen vaccine. Neuro-Oncology 2020. https://doi.org/10.1093/neuonc/noaa050.

[64] Calis JJA, Rosenberg BR. Characterizing immune repertoires by high throughput sequencing: strategies and applications. Trends Immunol 2014;35(12):581–90. https://doi.org/10.1016/j.it.2014.09.004.

[65] Heather JM, Ismail M, Oakes T, Chain B. High-throughput sequencing of the T-cell receptor repertoire: pitfalls and opportunities. Brief Bioinform 2018. https://doi.org/10.1093/bib/bbw138.

[66] Shugay M, Britanova OV, Merzlyak EM, et al. Towards error-free profiling of immune repertoires. Nat Methods 2014;11(6):653–5. https://doi.org/10.1038/nmeth.2960.

[67] Anon. Rapid amplification of 5′ complementary DNA ends (5′ RACE). Nat Methods 2005;2(8):629–30. https://doi.org/10.1038/nmeth0805-629.

[68] Li S, Lefranc MP, Miles JJ, et al. IMGT/HighV QUEST paradigm for T cell receptor IMGT clonotype diversity and next generation repertoire immunoprofiling. Nat Commun 2013;4. https://doi.org/10.1038/ncomms3333.

[69] Shugay M, Bagaev DV, Zvyagin IV, et al. VDJdb: a curated database of T-cell receptor sequences with known antigen specificity. Nucleic Acids Res 2018;46(D1):D419–27. https://doi.org/10.1093/nar/gkx760.

[70] Li B, Li T, Wang B, et al. Ultrasensitive detection of TCR hypervariable-region sequences in solid-tissue RNA-seq data. Nat Genet 2017;49(4):483–4. https://doi.org/10.1038/ng.3820.

[71] Zhigalova EA, Izosimova AI, Yuzhakova DV, et al. RNA-Seq-based TCR profiling reveals persistently increased intratumoral clonality in responders to anti-PD-1 therapy. Front Oncol 2020;10. https://doi.org/10.3389/fonc.2020.00385.

[72] Page DB, Yuan J, Redmond D, et al. Deep sequencing of T-cell receptor DNA as a biomarker of clonally expanded TILs in breast cancer after immunotherapy. Cancer Immunol Res 2016;4(10):835–44. https://doi.org/10.1158/2326-6066.CIR-16-0013.

[73] Sims JS, Grinshpun B, Feng Y, et al. Diversity and divergence of the glioma-infiltrating T-cell receptor repertoire. Proc Natl Acad Sci U S A 2016;113(25):E3529–37. https://doi.org/10.1073/pnas.1601012113.

[74] Melody SH, Shaina S, Tina W, et al. TCR sequencing can identify and track glioma-infiltrating t cells after dc vaccination. Cancer Immunol Res 2016;4(5):412–8. https://doi.org/10.1158/2326-6066.CIR-15-0240.

[75] Zhao J, Chen AX, Gartrell RD, et al. Immune and genomic correlates of response to anti-PD-1 immunotherapy in glioblastoma. Nat Med 2019;25(3):462–9. https://doi.org/10.1038/s41591-019-0349-y.

[76] Cloughesy TF, Mochizuki AY, Orpilla JR, et al. Neoadjuvant anti-PD-1 immunotherapy promotes a survival benefit with intratumoral and systemic immune responses in recurrent glioblastoma. Nat Med 2019;25(3):477–86. https://doi.org/10.1038/s41591-018-0337-7.
[77] Schalper KA, Rodriguez-Ruiz ME, Diez-Valle R, et al. Neoadjuvant nivolumab modifies the tumor immune microenvironment in resectable glioblastoma. Nat Med 2019;25(3):470–6. https://doi.org/10.1038/s41591-018-0339-5.
[78] Mohme M, Schliffke S, Maire CL, et al. Immunophenotyping of newly diagnosed and recurrent glioblastoma defines distinct immune exhaustion profiles in peripheral and tumor-infiltrating lymphocytes. Clin Cancer Res 2018;24(17):4187–200. https://doi.org/10.1158/1078-0432.CCR-17-2617.

CHAPTER

3

Heat shock protein vaccines in glioblastoma

Aden P. Haskell-Mendoza and Orin Bloch

Department of Neurological Surgery, University of California, Davis, Sacramento, CA, United States

Abstract

Heat shock proteins (HSPs) are cellular chaperones abundant in mammalian cells, and play an important role in response to stressors such as hyperthermia, hypoxia, and oxidative stress. The peptide-binding property of HSPs, together with their ability to deliver antigens to antigen-presenting cells, makes them a target for modulation in cancer immunotherapy. As a result, an autologous, patient-derived HSP-96 peptide complex vaccine (HSPPC-96) has been leveraged as an active immunotherapy to stimulate innate and adaptive responses in multiple cancers, including glioblastoma (GBM). To date, HSPPC-96 has demonstrated safety and tolerability in phase I and single-arm phase II trials for recurrent and newly diagnosed GBM, while producing robust immune responses as measured by increased leukocyte counts. Compared with historical controls, the single-arm phase II trials of HSPPC-96 for newly diagnosed and recurrent GBM demonstrated increased progression-free survival (PFS) and overall survival (OS), spurring a phase IIb randomized trial of the vaccine in combination with or in comparison with bevacizumab for recurrent GBM. The impact of immune checkpoints on responsiveness to the vaccine has been identified through post hoc analysis of completed clinical trials, and is now the target of a combined approach utilizing HSPPC-96 vaccination with immune checkpoint inhibition in newly diagnosed GBM.

Abbreviations

ALC	absolute leukocyte count
AMC	absolute monocyte count
APC	antigen-presenting cell
BBB	blood-brain barrier
CAR	chimeric antigen receptor
CNS	central nervous system
CSC	cancer stem cells
CTL	cytotoxic $CD8^+$ T lymphocyte
DC	dendritic cell
ER	endoplasmic reticulum

Immunotherapeutic Strategies for the Treatment of Glioma
https://doi.org/10.1016/B978-0-12-819755-4.00015-1

GBM	glioblastoma multiforme
HGG	high-grade glioma
HSSPC-96	heat shock protein peptide complex-gp96
NK	natural killer
OS	overall survival
PBL	peripheral blood leukocyte
PFS	progression-free survival
TAA	tumor-associated antigen
TIL	tumor-infiltrating lymphocyte
TSA	tumor-specific antigen
WBC	white blood cell count

Conflict of interest

No potential conflicts of interest were disclosed.

Introduction

Glioblastoma (GBM), the most common malignant primary tumor of the central nervous system (CNS), is an aggressive cancer with a poor prognosis [1]. Despite advances in surgery, chemotherapy, and radiation, GBM eventually becomes resistant to conventional therapy and is nearly always fatal, with a median overall survival (OS) of 16 months and a dismal 4.7% 5-year survival rate [2, 3]. This resistance to conventional therapy has led to an investigation into alternative treatment modalities, including extensive research into immunotherapy for GBM.

Immunotherapy, in its many forms, has demonstrated a significant benefit to patients with a number of solid cancers over the past decade. Immune targeting of tumor-specific antigens can minimize collateral damage to healthy tissue and lead to effective treatment with a more tolerable toxicity profile. In addition, some immunotherapy possesses the potential for long-term antitumor defense through sustained immunologic memory after discontinuation of the therapy. Unfortunately, despite the stunning success of immune checkpoint inhibition in multiple cancers, the efficacy for GBM has remained limited [4–6]. In part, this is due to the unique challenges of applying immunotherapy to the CNS.

To mount a successful antitumor immune response, a series of critical events must occur: (1) A tumor-specific, HLA-presented antigens must be present within the tumor and released, often due to cell death; (2) antigen-presenting cells (APCs) must uptake and present the antigen for T-cell priming and activation, usually within lymph nodes; (3) APCs and T cells must be sufficiently stimulated to result in clonal expansion of tumor-specific cytotoxic T lymphocyte (CTLs); (4) CTLs must traffic to, infiltrate, and recognize the tumor cells; (5) CTLs must have sufficient cytotoxic activity to kill the tumor and must resist local antiinflammatory signals [7]. In highly immunogenic tumors with spontaneous hematogenous and lymphatic metastasis, such as melanoma, most of these steps occur innately. Simply enhancing the innate immune response with immune checkpoint inhibitors (ICIs) has been highly successful in such tumors. This strategy, however, has failed in GBM, perhaps because antigen presentation and T-cell priming in the CNS are substantially different [8].

Immune activation in the CNS

Passage of immune cells in and out of the tissues of the CNS is limited by the blood-brain barrier (BBB), and most initial immune responses in the brain are instead performed by local antigen recognition through microglia and astrocytes; only following initiation of this response does the BBB allow the entry of immune cells from the periphery [9–11]. High-grade gliomas (HGGs) like GBM exhibit marked necrosis and inflammation as tumor growth outpaces vascular engraftment, and are infiltrated by peripheral immune cells [12, 13]. However, the majority of these cells are macrophages, with T cells making up only a small percentage [14]. Even upon infiltration, however, the majority of immune cells adopt or maintain immunosuppressive phenotypes, with large numbers of M2-polarized macrophages, as well as $CD4^+/CD25^+$/FOXP3-expressing regulatory T cells (T_{regs}) [15, 16]. The immunogenicity of the tumor is further limited by its moderate mutational burden relative to highly mutagenic and less cell-to-cell heterogenous cancers like melanoma and cancers of the lung [17, 18]. Given these factors, GBM is a representative example of an immunologically "cold" tumor. A biological explanation for these findings is the discovery of GBM-derived IL-6 induced increased expression of the PD-L1 (programmed death-ligand 1, B7H1) protein on the surface of peripheral monocytes and tumor-associated macrophages in patients with GBM, which mediates an immunosuppressive effect on infiltrating CTLs [19, 20]. Any immunotherapeutic attempts to treat these tumors will need to overcome these additional challenges and restore a more immunoreactive, "hot" phenotype.

Current immunotherapies can be broadly classed into active and passive approaches. Passive immunotherapies rely on transferring immunoreactive molecules, antibodies, or activated immune cells engineered exogenously into the patient to target tumor antigens [21]. Examples include immune checkpoint-modulating drugs such as pembrolizumab (anti-PD1) or ipilimumab (anti-CTLA-4); adoptive transfer of ex vivo expanded chimeric antigen receptor (CAR) T cells to attack based on tumor-specific antigens is another example. Passive immunotherapies are limited in that the response to therapy lasts only as long as the therapy is given, and checkpoint inhibitors, despite strong preclinical data, have shown a limited efficacy in glioblastoma [4, 6]. Given that the tumor and its microenvironment exhibit profound immunosuppression, the dependence of immune-modulating therapies on $CD8^+$ T-cell activation may make them poor candidates as single agents for GBM.

Vaccines for GBM

To overcome the challenges of immune stimulation for tumors in the CNS, antitumor vaccination can be performed to bypass the scavenging and presentation of antigens behind the BBB. This strategy has most commonly been implemented in the form of peptide, autologous tumor cell, or dendritic cell vaccines. Peptide vaccination is accomplished by delivering synthetic or tumor-derived short peptides meant to evoke a T-cell response through uptake and presentation by endogenous APCs. Such vaccines are easiest to scale up for mass production, but utility is limited by the ability to identify effective candidate peptide antigens [22]; single-peptide agents can easily be overcome via downregulation or negative immune selection of the targeted antigen, resulting in immune escape [23]. Polyvalent, patient-specific peptide vaccines require substantially more effort to identify and produce, negating the general benefits of a peptide approach. Autologous tumor cell vaccines may, in part, overcome this disadvantage by using lysates of resected tumors codelivered with adjuvant immune-stimulating agents, obviating the need for the identification of the specific antigens. Furthermore, delivery of a variety of antigenic peptides reduces the risk of immune escape. Despite these advantages, autologous tumor cell vaccines do not produce robust immune responses and are not in widespread use. They also require patient-specific tissue collection and processing, substantially increasing the cost compared with off-the-shelf peptide vaccines. Dendritic cell (DC) vaccines have been the subject of the most investigation for active immunotherapy approaches. Production of vaccine most commonly involves leukapheresis of the patient, followed by pulsation of the isolated DCs with tumor cell lysate, peptides, TSAs, or TAAs. Expanded DCs are then injected into the patient; by harnessing these professional antigen-presenting cells, MHC class I and MHC class II-mediated activation of $CD4^+$ helper T cells and $CD8^+$ cytotoxic T cells can be directly achieved [24]. However, this approach is the most labor-intensive and expensive, requiring a separate collection of leukocytes and tumor tissue (for autologous tumor-loaded DC vaccines), as well as expansion and preservation in a GMP facility. Proper collection of leukocytes for DC expansion is particularly difficult in GBM patients, as leukapheresis is most effective when patients are not leukopenic or on high doses of corticosteroids. GBM patients often require steroids at diagnosis of a tumor to control associated edema, which continues after surgery, and then become leukopenic as a result of adjuvant chemotherapy. This leaves a narrow window for effective leukapheresis between surgery and radiation.

Heat shock protein-peptide complexes (HSPPCs) are a subclass of peptide vaccines that combine the advantages of peptide, autologous-derived, and dendritic cell vaccines to produce robust immune responses in humans. Heat shock proteins (HSPs) act as molecular chaperones to salvage or guide the degradation of misfolded proteins resulting from cell stressors, such as hyperthermia, oxidative stress, hypoxia, radiation, and inflammation [25, 26]. The high level of abnormal proteins produced by cancers suggests a role for HSPs in chaperoning tumor peptides, some of which will be antigenic [27]. Extracellular HSPs are recognized by scavenger receptors on the surface of APCs and can guide their chaperoned peptides to MHC class I and II presentation [28]. HSPs also activate toll-like receptors on APCs, releasing proinflammatory cytokines that assist with T-cell activation [29]. By harvesting HSPPCs from autologous tumor cells, both an innate and adaptive immune response can be elicited. A personalized, polyvalent immune activation is initiated, without the need for costimulatory agents and with significantly less effort than DC vaccines. In this chapter, we provide an overview of the biological function of HSPs and how HSPPCs have been utilized clinically in the setting of GBM.

Heat shock proteins in health and disease

Within cells, heat shock proteins typically act as molecular chaperones as part of the normal stress response [26]. HSPs are highly conserved across all domains of life and presently are known to exist as many distinct protein families within the human genome. Initially named according to molecular weight, major HSP families include the small HSPs, HSP60, HSP70, HSP90, and HSP100 [28]. The high level of duplication of the genes in several protein families is indicative of a varied subcellular localization and functional specialization. Intracellularly, HSPs' canonical chaperone function is to aid in the stabilization, folding, and transport of newly formed proteins, particularly when cellular stressors have resulted in misfolding or aggregation; they additionally aid in the clearance and degradation of unsalvageable proteins. Under normal conditions, this process results in the restoration of proper biological function for salvaged proteins, often in a stepwise manner as misfolded proteins are passed from HSP to HSP. The production of HSP transcripts and polypeptides in response to cell stress is finely regulated by the transcription factor heat shock factor 1 (HSF1). The transcriptional and translational dysregulation that occur in the setting of cancer, in particular due to the increased levels of aberrant protein products, are posited as a mechanistic driver for the high HSP levels seen in tumor cells [30].

In addition to their intracellular function, several HSPs, including HSP70, HSP90, calreticulin, and HSP gp96, have been shown to have immune function as membrane-bound or extracellular proteins [28, 31]. In in vitro and in vivo models of glioblastoma, constitutively high expression of αB-crystallin, HSP27, HSP70, and HSP90 has been shown to be present prior to heat induction experiments, possibly indicating an uncoupling of protein expression from the "normal" stress response in this cancer [32]. There is also evidence of secretion into the microenvironment via the packaging of HSPs on exosomes released by GBM cells [33, 34]. Beyond binding of misfolded proteins or aggregates, several HSPs, including the HSP70 and HSP90 families, have been shown to bind small peptides that are potentially antigenic and able to elicit immune clearance of the tumor [31]. In particular, the HSP90 family is responsible for loading such peptides onto MHC class I molecules, and interacts as a chaperone or co-chaperone with protein members of key genetic pathways involved in malignant transformation and tumor growth such as p53, AKT, CDK4, MAPK, EGFR, TGFβ, and mTOR [35]. Based on these findings, HSPs have been leveraged as carriers for antigenic peptides to activate the immune system in cancer.

Heat shock protein-peptide complexes

A key driver for the investigation of HSPs as potential therapeutics is their ability to stimulate innate and adaptive immunity. While classically, exogenous antigens are presented by APCs on MHC class II molecules for recognition by $CD4^+$ helper T cells, HSPs are capable of cross-presenting antigens by using the class II and endogenous MHC class I pathway.

The therapeutic potential of HSPs as an immunogenic agent was first brought to light by animal studies showing that HSP gp96, an endoplasmic reticulum (ER)-associated variant of the HSP90 family, could load antigens onto the endogenous (MHC class I) antigen recognition pathway resulting in $CD8^+$ cytotoxic T-lymphocyte activation, and that treatment with heat shock protein preparations derived from autologous cancer cells could reduce tumor growth and metastasis [36, 37]. Neither the HSP alone nor the associated peptides were sufficient to induce an antitumor response; instead, the antitumor inductive fraction was demonstrated to consist of HSP-peptide complexes, or HSPPCs [38, 39]. Subsequent studies have shown that the internalization of HSPPCs occurs in part via interaction of the complex with receptors on

APCs, such as CD91, CD14, LOX-1, TLR-2, and TLR-4 [29]. Following internalization, HSPPCs can follow a cytosolic pathway consisting of proteasomal processing, transportation to the ER, and loading onto MHC class I molecules; independently of the proteasome, HSP complexes can be loaded onto MHC class I and class II molecules via the receptor-mediated endosomal pathway [40–42] (Fig. 1).

Beyond adaptive stimulatory activity, HSPs have additional effects on the innate component of the immune system by interacting with cell surface receptors to stimulate pathways synapsing on NF-κB activation in APCs. Proinflammatory cytokines, including granulocyte macrophage colony-stimulating factor (GM-CSF), IL-12, IL-1β, and TNF-α, are produced following the interaction of HSPPCs with macrophages. Inducible nitric oxide synthase (iNOS) is also activated in macrophages and dendritic cells following exposure to HSPPCs. These signals potentiate the activation of both CTLs and the natural killer (NK) cells of the innate immune system and promote maturation of dendritic cells; together with the polarization of macrophages toward an M1 phenotype, HSPPCs create a general proinflammatory response [29]. These promising preclinical data have been key motivators for a number of phase I, II, and III clinical trials of HSPPC vaccines for cancer.

In addition to their ability to stimulate adaptive and innate immunity, HSPPCs present the additional advantage of not being limited to a predetermined antigen that can be overcome by the tumor as it evolves under immune pressure. As the majority of GBMs exhibit a great deal of genomic, transcriptional, and translational heterogeneity from cell to cell within the tumor, deriving HSPPCs directly from the patient tumor provides a polyvalent, patient-specific vaccine that is better able to educate the immune system and more likely to stimulate a response against the tumor [43, 44]. While it is true that vaccinating with tumor-derived HSPPCs consisting of unknown antigens may target nonessential antigens to tumor survival, or antigens expressed by a subpopulation of tumor cells, spreading of antigenic epitopes among immune cells activated by the vaccine may hypothetically produce a broader response, further increasing the epitope coverage of the vaccine and allowing better targeting of varied and dynamic tumor cells.

The HSPPC product that has been most deeply investigated in clinical trials is heat shock protein-peptide complex 96 (HSPPC-96; Prophage, Agenus Inc.), consisting of patient-autologous peptides and the chaperone HSP glycoprotein-96 (HSP gp96), the endoplasmic reticulum-associated member of the HSP90 family that has been shown to be secreted extracellularly and activate such receptors as CD91 and TLR2/4. Though other HSP families have similar activity, pilot studies of HSPPC-96 have shown the vaccine to have a favorable toxicity and safety profile, and well suited to manufacture as a clinical grade product. Following resection, patient tumor is frozen and shipped to the manufacturing site, where HSPs are purified via liquid chromatography isolation. Following isolation, HSPs are subjected to denaturing gel electrophoresis and western blotting with anti-gp 96 antibody. Additional measures, including quality control and sterility screening, result in delivery of a completed vaccine 3–4 weeks following tumor resection; dosing schedules typically consist of weekly vaccination during the first 4 weeks of therapy, shifting to biweekly treatment until the depletion of vaccine supply [45]. This process is more intensive than generating a generic peptide vaccine based on the sequence of one or a few generalized epitopes. However, the benefits of a personalized and polyvalent response justify the effort and cost differential. In comparison with DC vaccines loaded with autologous tumor products, HSP vaccines are significantly less

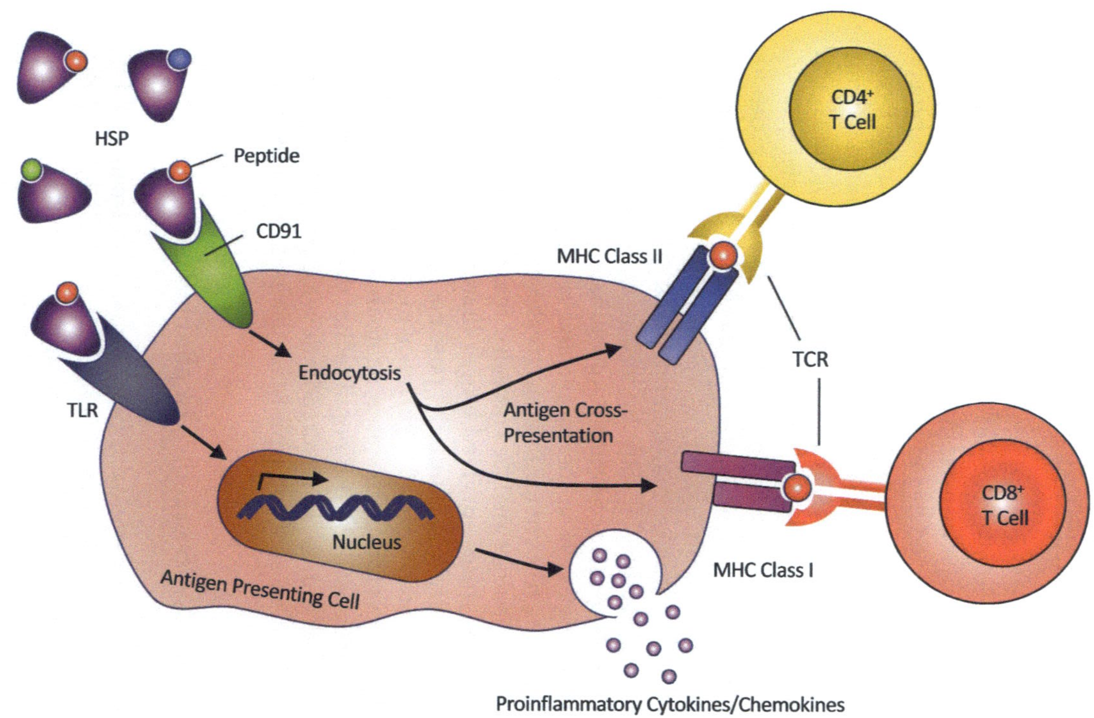

FIG. 1 Immune activation by heat shock protein-peptide complexes. Peptides mounted on HSPs can be endocytosed into APCs through the activation of the CD91 receptor, leading to peptide cleaving and cross-presentation on MHC class I and class II, to stimulate T-cell receptors and induce clonal expansion of $CD4^+$ and $CD8^+$ T cells. Binding of HSPs to surface toll-like receptors (TLR2/4) stimulates NF-κB activation resulting in proinflammatory cytokine production.

expensive and time-consuming to produce. The ability to generate HSP vaccines is also not dependent on the peripheral immune function (leukopenia) of the patient at the time of generation.

Safety and efficacy of HSPPC-96 vaccines

The HSPPC-96 vaccine has been leveraged in a variety of cancers to date in attempts to elicit a patient-specific immune response. A pilot study of HSPPC-96 in an initial cohort of 16 patients suffering from a variety of advanced malignancies, including stage IV breast, pancreatic, and colon cancer, was conducted by Janetzki et al. to establish the general safety profile of the vaccine and measure immune activation against the tumor by T-cell Elispot and immune monitoring. In this study, 6/12 immunized patients generated tumor-specific $CD8^+$ CTLs; additionally, expansion of NK cells was seen in 8/13 patients. Overall, severe toxicity or autoimmunity due to vaccination was not observed. The trial generally validated preclinical data and provided a robust foundation for further investigation despite limitations in sample size and design preventing clear evaluation of efficacy [46]. Additional phase I/II studies in a variety of malignancies, including colorectal cancer, melanoma, nonsmall-cell lung cancer, and pancreatic adenocarcinoma, demonstrated promising safety, in addition to practicality of vaccine manufacture [47–50]. Interestingly, several trials noted signs of tumor-specific immune response postvaccination. Unfortunately, two separate randomized, multicenter phase III trials of HSPPC-96 in melanoma and renal cell carcinoma did not find differences in overall survival or recurrence-free survival rate relative to controls [51, 52]. However, in the melanoma trial, subset analyses revealed that those with early-stage disease and a greater number of vaccine doses survived longer than those with fewer doses, when controlling for bias due to longevity [51]. Similarly, post hoc analyses of the renal cell carcinoma trial found improved PFS and reduced recurrence in a subgroup of intermediate-risk patients risk stratified by ECOG performance status, again indicating an increased benefit for HSPPC-96 in earlier stages of disease [53]. It is possible that this benefit arises due to the fact that later-stage, more advanced tumors have already developed multiple complex mechanisms to evade the immune system as they begin to metastasize from their tissue of origin.

HSPPC-96 vaccines in glioblastoma

Following trials in other malignancies establishing the safety of treatment, HSPPC-96 vaccination was applied as a therapy for recurrent GBM. Crane et al. reported the results of a phase I clinical trial of autologous HSPPC-96 vaccination for recurrent high-grade gliomas [54]. Twelve enrolled patients met inclusion criteria following resection of tumor. Patients received 25 μg doses of HSPPC-96 weekly or biweekly for the first four treatments. Following the receipt of four vaccinations, all patients received 25 μg of vaccine biweekly until the depletion of vaccine. Peripheral blood leukocytes (PBLs) and tumor-infiltrating lymphocytes (TILs) were isolated for immune monitoring and phenotyping by flow cytometry and various in vitro assays, including stimulation of PBLs with leftover autologous vaccine and qRT-PCR. 11/12 (92%) patients had significant immune responses to treatment with HSPPC-96.

The study demonstrated increased production and secretion of the cytokine interferon gamma (IFN-γ) and expansion of T cells in ex vivo PBLs restimulated with autologous HSPPC-96 in a manner that correlated with a proinflammatory clinical response. Tumors resected from patients following treatment with HSPCC-96 displayed increased infiltrating IFN-γ-producing T cells and NK cells; interestingly, $CD4^+$, $CD25^+$, and $FOXP3^+$ T_{regs} were decreased in patient peripheral blood. The 11 immune responders had a median OS of 47 weeks, compared with 16 weeks for the lone nonresponder. Overall, the study demonstrated the feasibility of autologous HSPPC-96 production in recurrent GBM and produced compelling evidence of the immunogenic effect of the vaccine [54].

A subsequent multicenter phase II, single-arm trial of HSPPC-96 for recurrent glioblastoma was reported by Bloch et al. [55]. Forty-one patients met the inclusion criteria for the study and were treated with tumor-derived HSPPC-96 vaccine, 25 μg doses every week for 4 weeks followed by biweekly vaccination until the depletion of vaccine or tumor progression. Median PFS was 19.1 weeks and median OS was 42.6 weeks, with a 6-month OS of 90.2% and a 12-month OS of 29.3%. Pretreatment immune status was assessed by blood sampling prior to surgery. Complete blood counts with differentials of white blood cell count (WBC), absolute lymphocyte count (ALC), and absolute monocyte count (AMC) were obtained. The study found that preoperative ALC greater than or equal to the median of 0.9×10^9 cells/L significantly improved overall survival as compared with patients with ALC less than median (49.1 vs 37.1 wk). Additionally, ALC was found to be a predictor of decreased survival outcome in a proportional hazards model, with a relative hazard ratio of 4.0 for patients with ALC less than median; AMC and WBC did not predict outcome. The majority, 27/41 (66%), of patients were lymphopenic at preoperative blood draw according to the standard lower limit of 1.0×10^9 cells/L. Due to the single-arm study design, comparisons with historical controls were leveraged to evaluate outcomes. The study results favorably compared with the results of a similarly selected patient group in the PRECISE trial, a phase III trial comparing carmustine wafers implanted into the resection site vs convection-enhanced delivery of a chimeric IL-13-pseudomonas exotoxin fusion protein. The study authors of the PRECISE trial did not achieve a significant difference in survival between treatments, with a median OS of 36.4 weeks in the fusion protein group vs 35.3 weeks in the carmustine wafer-treated patients. Relative to these survival outcomes in a large clinical trial in recurrent GBM, the reported median OS of 42.6 weeks for patients treated with HSPPC-96 was seen as an encouraging result for further investigation of the vaccine in multiarm clinical trials [55].

Although there is no standard of care for recurrent GBM, the anti-VEGF monoclonal antibody, bevacizumab, is commonly used to treat recurrent disease. Biologically, the drug inhibits angiogenesis by binding to the vascular endothelial growth factor A (VEGF-A) protein [56]. Following up the single-arm phase II study of HSPPC-96, the Alliance for Clinical Trials in Oncology A071101 trial was developed as a randomized, multicenter phase II study of the efficacy of HSPPC-96 alone and in combination with bevacizumab vs bevacizumab alone for recurrent GBM. Ninety patients met the inclusion criteria for the study and were randomized to receive either: HSPPC-96 vaccine until tumor progression, at which point bevacizumab was initiated; concurrent HSPPC-96 and bevacizumab; or bevacizumab alone, with a primary study endpoint of overall survival. Patients were enrolled at a distribution of 29:30:31 between these treatment groups. While the vaccine was again demonstrated to be safe and well tolerated, unfortunately, OS for the combined HSPPC-96 groups was 7.5 months vs

10.7 months for bevacizumab alone, and the study was terminated at interim analysis for futility due to the lack of demonstrable survival benefit [57].

Both the phase I and phase II single-arm trials of HSPPC-96 in recurrent GBM showed that immune function, either as measured ex vivo and intratumorally (Crane et al.), or through a surrogate marker of ALC (Bloch et al.), correlated with the patient's survival. Therefore, pretreatment lymphopenia and a generally immunosuppressed state induced by GBM is an important factor that warrants evaluation by the clinician prior to treatment. As discussed, GBM-derived cytokines are capable of increasing PD-L1 expression on peripheral and tumor-infiltrating macrophages. A focus on targeting checkpoint-associated immune dysfunction in combination with vaccination may increase the clinical benefit to patients.

HSPPC-96 has also been investigated in newly diagnosed GBM. In a single-arm, phase II study, 46 patients were enrolled and received fractionated radiotherapy to a dose of 60 Gy (2 Gy fractions × 30) and oral temozolomide (75 mg/m^2); 2–5 weeks following radiotherapy, 25 μg doses of vaccine was administered weekly for 4 weeks. Maintenance temozolomide therapy was initiated following the administration of vaccine, with a fifth dose of vaccine administered at the initiation of therapy. During maintenance temozolomide, vaccine, as available, was administered initially 3 weeks after the fifth dose, followed by monthly administrations subsequently. Patients were subjected to immunologic evaluation, with peripheral blood obtained at the time of craniotomy for the isolation of PBLs. Patients received vaccine until depletion or disease progression, and were evaluated for OS and PFS. Median PFS was 18.0 months, and median OS was 23.8 months. In the study, both MGMT promoter methylation and low PD-L1 expression on peripheral myeloid cells were found to be favorable prognostic factors for increased overall survival (MGMT methylated vs unmethylated: 44.7 months vs 18.0 months; PD-L1 low vs high: 44.7 months vs 18.0 months). In a multivariate proportional hazards model of overall survival in this patient cohort, MGMT methylation status, myeloid PD-L1 expression, and Karnofsky performance status (KPS) were demonstrated to independently predict the overall survival. The data found in this study were compared with the results of patients in the ACT IV phase II trial of rindopepimut for newly diagnosed GBM on the basis of their similarly strict inclusion criteria. Rindopepimut is a peptide vaccine targeting the EGFRvIII neoantigen expressed in roughly 30% of glioblastomas; it consists of a 13-amino acid peptide derived from EGFRvIII conjugated to keyhole limpet hemocyanin. 405 patients included in the ACT IV study were found to have a median OS of 20.1 months vs 20.0 months for the rindopepimut-vaccinated vs control arms. The 23.8 median OS in the HSPPC-96 study for newly diagnosed GBM compares favorably to this large study with similar patients. However, it is important to note that while the PFS of 18 months following treatment with HSPPC-96 compares favorably to similar trials, the median survival following progression remained roughly 6 months, a phenomenon seen in multiple other studies of recurrent GBM. In the case of HSPPC-96, this initial period of PFS may be derived from the activation of the immune system as stimulated by the vaccine and subsequent control of the tumor; however, the heterogenous tumor likely circumvents immune surveillance as cells without neoantigens presented by the vaccine are selected for and/or via engaging additional mechanisms of immunoresistance. Encouragingly, this study did suggest a benefit for HSPPC-96 in a single-arm phase II trial and additionally demonstrated an important, independent role for PD-L1 status in the prognosis of newly diagnosed GBM [58]. The results of all of the clinical trials of HSPPC-96 for GBM are summarized in Table 1.

TABLE 1 Clinical trials of HSPPC-96 in adult glioblastoma.

Citation	Year	ClinicalTrials.gov identifier	Phase	*n*	Newly diagnosed or recurrent	Comparator	Median OS	Immunologic effects
Crane et al. [54]	2012	NCT00293423	I	12	R	N/A	47 weeks (responders)	Ex vivo immune responses in PBL following vaccination, increases in tumor-infiltrating T and NK cells
Bloch et al. A [55]	2014	NCT00293423	II	41	R	Single-arm; compared with historical controls	42.6 weeks (95% CI: 34.7–50.5)	Patients with absolute lymphocyte count below cohort median had decreased OS
Bloch et al. B [58]	2017	NCT00905060	II	46	ND	Single-arm; compared with historical controls	23.8 months (95% CI: 19.8–30.2)	Patients with high PD-L1 expression in peripheral myeloid cells had decreased OS
Bloch et al. C [57]	2017	NCT01814813	II	90	R	BEV vs HSPPC-96 + concurrent BEV vs HSPPC-96 + BEV at progression	7.5 months (HSPPC-96 treated groups) vs 10.7 months (BEV alone)	N/A
Not yet published	N/A	NCT03018288	II	108 (planned)	ND	SOC + Pembro vs SOC + Pembro + Placebo Vaccine vs SOC + Pembro + HSPPC-96	N/A	N/A

ND, newly diagnosed; *R*, recurrent; *BEV*, bevacizumab; *SOC*, standard of care; *Pembro*, pembrolizumab; *OS*, overall survival; *PBL*, peripheral blood leukocytes.

The PD-L1 surface protein is capable of inducing T-cell anergy and apoptosis by binding to the PD-1 receptor-activated T cells [59]. This natural protection mechanism against autoimmunity is harnessed by GBM through increased expression of PD-L1 on the tumor cell surface, as well as the induction of PD-L1 on circulating myeloid cells through the production and secretion of tumor-derived cytokines [19, 20, 60]. However, studies have shown that there is no correlation between *tumor* PD-L1 status and clinical outcome, which may indicate the necessity of increased focus on addressing the role of myeloid PD-L1 in GBM immunoresistance [60]. Furthermore, despite the efficacy of several FDA-approved anti-PD-1 and anti-PD-L1 monoclonal antibody drugs (immune checkpoint inhibitors) against other cancers, several large trials in GBM have not shown clinical benefit for these therapies [4, 6]. As HSPPC-96 is demonstrably capable of reeducating the immune system to produce an antitumor response, there may be a role for checkpoint inhibitors in supplementing the activity of HSPPC-96 by passively reducing tumoral immunosuppression. At the time of this writing, a randomized, multicenter phase II clinical trial of radiation therapy, temozolomide, and the anti-PD-1 drug pembrolizumab with or without HSPPC-96 in newly diagnosed GBM has been initiated, with a planned enrollment of 108 patients (NCT03018288). This study will provide valuable information as to the clinical value of the addition of active immunotherapy by reeducating the immune system while attempting to address the systemic immunosuppression mediated by GBM with checkpoint inhibitors. As HSPPC-96 has previously been shown to be well tolerated with few high-grade immune-related adverse effects, combination with pembrolizumab may prove better tolerated than the autoimmunity observed in the setting of combination therapy with multiple checkpoint inhibitors.

Conclusion

Immunotherapy for the treatment of cancer remains a promising field with many future avenues for investigation. The immunologically cold nature of glioblastoma necessitates that future therapies exploit a variety of mechanisms to elicit an antitumor response. In preclinical studies, heat shock protein-peptide complex vaccines have been shown to be highly effective at cross-presenting antigens on class I and II MHC, and stimulating proinflammatory cytokine production through receptor activation. Theoretically, HSP-based cancer vaccines have an advantage over other peptide vaccines because they do not require predetermination of a specific tumor antigen, and instead deliver a polyvalent, personalized immunogen. Additionally, the innate proinflammatory nature of heat shock proteins may obviate the need for immune-stimulating adjuvants. Compared with tumor lysate-pulsed dendritic cell vaccines, heat shock protein vaccines may be expected to deliver equivalent efficacy with substantially less effort and cost to generate an autologous product.

Clinically, HSPPC-96 has been shown to be safe and well tolerated in many cancer types, including GBM. In GBM, as opposed to other cancer types, HSPPC-96 has also been shown to reliably induce an immune response. Due to limitations in antigen presentation from behind the BBB, vaccination with agents such as HSPPC-96 can play an important role in priming the immune system against tumor antigens. Although outcomes for patients treated with HSPPC-96 vaccination as a monotherapy have not been overwhelmingly positive in controlled trials, subgroup analysis demonstrates that peripheral immune checkpoint expression, in particular PD-L1, significantly impacts the outcome of vaccine treatment. As a result, combination therapy utilizing the vaccine with immune checkpoint inhibition aims to demonstrate further improvement in patient survival.

References

[1] Omuro A, DeAngelis LM. Glioblastoma and other malignant gliomas: a clinical review. JAMA 2013;310(17):1842–50. https://doi.org/10.1001/jama.2013.280319.
[2] Gilbert MR, Dignam JJ, Armstrong TS, et al. A randomized trial of bevacizumab for newly diagnosed glioblastoma. N Engl J Med 2014;370(8):699–708. https://doi.org/10.1056/NEJMoa1308573.
[3] Ostrom QT, Gittleman H, Farah P, et al. CBTRUS statistical report: primary brain and central nervous system tumors diagnosed in the United States in 2006-2010. Neuro-Oncology 2013;15:1–56. https://doi.org/10.1093/neuonc/not151.
[4] Bristol-Myers Squibb announces phase 3 CheckMate -498 study did not meet primary endpoint of overall survival with opdivo (nivolumab) plus radiation in patients with newly diagnosed MGMT-unmethylated glioblastoma multiforme. Bristol-Meyers Squibb.
[5] Kurz SC, Cabrera LP, Hastie D, et al. PD-1 inhibition has only limited clinical benefit in patients with recurrent high-grade glioma. Neurology 2018;91(14):E1355–9. https://doi.org/10.1212/WNL.0000000000006283.
[6] Omuro A, Vlahovic G, Lim M, et al. Nivolumab with or without ipilimumab in patients with recurrent glioblastoma: results from exploratory phase i cohorts of CheckMate 143. Neuro-Oncology 2018;20(5):674–86. https://doi.org/10.1093/neuonc/nox208.
[7] Chen DS, Mellman I. Oncology meets immunology: the cancer-immunity cycle. Immunity 2013. https://doi.org/10.1016/j.immuni.2013.07.012.
[8] Desai K, Hubben A, Ahluwalia M. The role of checkpoint inhibitors in glioblastomaaa. Target Oncol 2019. https://doi.org/10.1007/s11523-019-00655-3.
[9] Carpentier PA, Begolka WS, Olson JK, Elhofy A, Karpus WJ, Miller SD. Differential activation of astrocytes by innate and adaptive immune stimuli. Glia 2005;49(3):360–74. https://doi.org/10.1002/glia.20117.
[10] Dong Y, Benveniste EN. Immune function of astrocytes. Glia 2001;36(2):180–90. https://doi.org/10.1002/glia.1107.
[11] Pachter JS, De Vries HE, Fabry Z. The blood-brain barrier and its role in immune privilege in the central nervous system. J Neuropathol Exp Neurol 2003;62(6):593–604. https://doi.org/10.1093/jnen/62.6.593.
[12] Giometto B, Bozza F, Faresin F, Alessio L, Mingrino S, Tavolato B. Immune infiltrates and cytokines in gliomas. Acta Neurochir 1996;138(1):50–6. https://doi.org/10.1007/BF01411724.
[13] Raza SM, Lang F, Aggarwal B, Fuller G, Wildrick D, Sawaya R. Necrosis and glioblastoma: a friend or a foe? A review and a hypothesis. Neurosurgery 2002;51(1):2–13. https://doi.org/10.1227/01.NEU.0000017462.65730.06.
[14] Parney IF, Waldron JS, Parsa AT. Flow cytometry and in vitro analysis of human glioma-associated macrophages: laboratory investigation. J Neurosurg 2009;110(3):572–82. https://doi.org/10.3171/2008.7.JNS08475.
[15] Saha D, Martuza R, Rabkin SD. Macrophage polarization contributes to glioblastoma eradication by combination immunovirotherapy and immune checkpoint blockade. Cancer Cell 2017;32(2):253–67. https://doi.org/10.1016/j.physbeh.2017.03.040.
[16] Hussain SF, Yang D, Suki D, Aldape K, Grimm E, Heimberger AB. The role of human glioma-infiltrating microglia/macrophages in mediating antitumor immune responses1. Neuro-Oncology 2006;8(3):261–79. https://doi.org/10.1215/15228517-2006-008.
[17] Alexandrov LB, Nik-Zainal S, Wedge DC, et al. Signatures of mutational processes in human cancer. Nature 2013;500(7463):415–21. https://doi.org/10.1038/nature12477.
[18] Samstein RM, Lee CH, Shoushtari AN, et al. Tumor mutational load predicts survival after immunotherapy across multiple cancer types. Nat Genet 2019;51(2):202–6. https://doi.org/10.1038/s41588-018-0312-8.
[19] Bloch O, Crane CA, Kaur R, Safaee M, Rutkowski MJ, Parsa AT. Gliomas promote immunosuppression through induction of B7-H1 expression in tumor-associated macrophages. Clin Cancer Res 2013;19(12):3165–75. https://doi.org/10.1158/1078-0432.CCR-12-3314.

[20] Lamano JB, Lamano JB, Li YD, et al. Glioblastoma-derived IL6 induces immunosuppressive peripheral myeloid cell PD-L1 and promotes tumor growth. Clin Cancer Res 2019;25(12):3643–57. https://doi.org/10.1158/1078-0432.CCR-18-2402.
[21] Mellman I, Coukos G, Dranoff G. Cancer immunotherapy comes of age. Nature 2011;480(22):480–9. https://doi.org/10.1038/nature10673.
[22] Purcell AW, McCluskey J, Rossjohn J. More than one reason to rethink the use of peptides in vaccine design. Nat Rev Drug Discov 2007;6(5):404–14. https://doi.org/10.1038/nrd2224.
[23] Weller M, Butowski N, Tran DD, et al. Rindopepimut with temozolomide for patients with newly diagnosed, EGFRvIII-expressing glioblastoma (ACT IV): a randomised, double-blind, international phase 3 trial. Lancet Oncol 2017;18(10):1373–85. https://doi.org/10.1016/S1470-2045(17)30517-X.
[24] Palucka K, Banchereau J. Cancer immunotherapy via dendritic cells. Nat Rev Cancer 2013;12(4):265–77. https://doi.org/10.1038/nrc3258.Cancer.
[25] Young JC, Agashe VR, Siegers K, Hartl FU. Pathways of chaperone-mediated protein folding in the cytosol. Nat Rev Mol Cell Biol 2004;5(10):781–91. https://doi.org/10.1038/nrm1492.
[26] Craig EA, Weissman JS, Horwich AL. Heat shock proteins and molecular chaperones: mediators of protein conformation and turnover in the cell. Cell 1994;78(3):365–72. https://doi.org/10.1016/0092-8674(94)90416-2.
[27] Ciocca DR, Calderwood SK. Heat shock proteins in cancer : diagnostic, prognostic, predictive, and treatment implications. Targeting heat shock proteins in cancer. Heat shock proteins in obesity: links to cardiovascular disease. Cell Stress Chaperones 2005;10(2):86–103.
[28] Schmitt E, Gehrmann M, Brunet M, Multhoff G, Garrido C. Intracellular and extracellular functions of heat shock proteins: repercussions in cancer therapy. J Leukoc Biol 2007;81(1):15–27. https://doi.org/10.1189/jlb.0306167.
[29] Srivastava P. Roles of heat-shock proteins in innate and adaptive immunity. Nat Rev Immunol 2002;2(3):185–94. https://doi.org/10.1038/nri749.
[30] Calderwood SK. Heat shock proteins and cancer: intracellular chaperones or extracellular signalling ligands? Philos Trans R Soc B Biol Sci 2018;373(1738). https://doi.org/10.1098/rstb.2016.0524.
[31] Binder RJ. Functions of heat shock proteins in pathways of the innate and adaptive immune system. J Immunol 2014;193(12):5765–71. https://doi.org/10.4049/jimmunol.1401417.
[32] Hermisson M, Strik H, Rieger J, Dichgans J, Meyermann R, Weller M. Expression and functional activity of heat shock proteins in human glioblastoma multiforme. Neurology 2000;1357–66.
[33] Graner MW, Alzate O, Dechkovskaia AM, et al. Proteomic and immunologic analyses of brain tumor exosomes. FASEB J 2009;23(5):1541–57. https://doi.org/10.1096/fj.08-122184.
[34] Chun S, Ahn S, Yeom CH, Park S. Exosome proteome of U-87MG glioblastoma cells. Biology (Basel) 2016;5(4):1–11. https://doi.org/10.3390/biology5040050.
[35] Graner MW, Bigner DD. Chaperone proteins and brain tumors: potential targets and possible therapeutics. Neuro-Oncology 2005;7(3):260–78. https://doi.org/10.1215/s1152851704001188.
[36] Suto R, Srivastava PK. A mechanism for the specific immunogenicity of heat shock protein-chaperoned peptides. Science 1995;269(5230):1585–8. https://doi.org/10.1126/science.7545313.
[37] Tamura Y, Peng P, Liu K, Daou M, Srivastava PK. Immunotherapy of tumors with autologous tumor-derived heat shock protein preparations. Science (80-) 1997;278(5335):117–20. https://doi.org/10.1126/science.278.5335.117.
[38] Srivastava PK, DeLeo AB, Old LJ. Tumor rejection antigens of chemically induced sarcomas of inbred mice. Proc Natl Acad Sci U S A 1986;83(10):3407–11. https://doi.org/10.1073/pnas.83.10.3407.
[39] Blachere NE, Li Z, Chandawarkar RY, et al. Heat shock protein-peptide complexes, reconstituted in vitro, elicit peptide-specific cytotoxic T lymphocyte response and tumor immunity. J Exp Med 1997;186(8):1315–22. https://doi.org/10.1084/jem.186.8.1315.
[40] Basu S, Binder RJ, Ramalingam T, Srivastava PK. CD91 Is a Common Receptor for Heat Shock Proteins gp96, hsp90, hsp70, and Calreticulin recently identified the 2 macroglobulin receptor CD91 as the receptor for gp96 (Binder et al., 2000a); gp96 molecules bind directly to CD91, and antibodies to CD91 inhi. Immunity 2001;14:303–13.
[41] Castellino F, Boucher PE, Eichelberg K, et al. Receptor-mediated uptake of antigen/heat shock protein complexes results in major histocompatibility complex class I antigen presentation via two distinct processing pathways. J Exp Med 2000;191(11):1957–64. https://doi.org/10.1084/jem.191.11.1957.
[42] Matsutake T, Sawamura T, Srivastava PK. High efficiency CD91- and LOX-1-mediated re-presentation of gp96-chaperoned peptides by MHC II molecules. Cancer Immun 2010;10(May):1–7.

[43] Verhaak RGW, Hoadley KA, Purdom E, et al. An integrated genomic analysis identifies clinically relevant subtypes of glioblastoma characterized by abnormalities in PDGFRA, IDH1, EGFR and NF1. Cancer Cell 2010;17 (1). https://doi.org/10.1016/j.ccr.2009.12.020.An.

[44] Neftel C, Laffy J, Filbin MG, et al. An integrative model of cellular states, plasticity, and genetics for glioblastoma. Cell 2019;178(4):835–49. https://doi.org/10.1016/j.cell.2019.06.024.

[45] Ampie L, Choy W, Lamano JB, Fakurnejad S, Bloch O, Parsa AT. Heat shock protein vaccines against glioblastoma: from bench to bedside. J Neuro-Oncol 2015;123(3):441–8. https://doi.org/10.1007/s11060-015-1837-7.

[46] Janetzki S, Palla D, Rosenhauer V, Lochs H, Lewis JJ, Srivastava PK. Immunization of cancer patients with autologous cancer-derived heat shock protein gp96 preparations: a pilot study. Int J Cancer 2000;88(2):232–8. https://doi.org/10.1002/1097-0215(20001015)88:2<232::AID-IJC14>3.0.CO;2-8.

[47] Mazzaferro V, Coppa J, Carrabba MG, et al. Vaccination with autologous tumor-derived heat-shock protein Gp96 after liver resection for metastatic colorectal cancer. Clin Cancer Res 2003;9(9):3235–45.

[48] Pilla L, Patuzzo R, Rivoltini L, et al. A phase II trial of vaccination with autologous, tumor-derived heat-shock protein peptide complexes Gp96, in combination with GM-CSF and interferon-α in metastatic melanoma patients. Cancer Immunol Immunother 2006;55(8):958–68. https://doi.org/10.1007/s00262-005-0084-8.

[49] Santis G, Senzer NN, Champagne P, Isakov L, Teofilovici F. Phase II feasibility study of autologous vaccine (HSPPC-96) in patients with resectable lung cancer. J Clin Oncol 2008;26(15 suppl).

[50] Maki RG, Livingston PO, Lewis JJ, et al. A phase I pilot study of autologous heat shock protein vaccine HSPPC-96 in patients with resected pancreatic adenocarcinoma. Dig Dis Sci 2007;52(8):1964–72. https://doi.org/10.1007/s10620-006-9205-2.

[51] Testori A, Richards J, Whitman E, et al. Phase III comparison of vitespen, an autologous tumor-derived heat shock protein gp96 peptide complex vaccine, with physician's choice of treatment for stage IV melanoma: the C-100-21 study group. J Clin Oncol 2008;26(6):955–62. https://doi.org/10.1200/JCO.2007.11.9941.

[52] Wood C, Srivastava P, Bukowski R, et al. An adjuvant autologous therapeutic vaccine (HSPPC-96; vitespen) versus observation alone for patients at high risk of recurrence after nephrectomy for renal cell carcinoma: a multicentre, open-label, randomised phase III trial. Lancet 2008;372(9633):145–54. https://doi.org/10.1016/S0140-6736(08)60697-2.

[53] Messing EM, Manola J, Wilding G, et al. Phase III study of interferon alfa-NL as adjuvant treatment for resectable renal cell carcinoma: an eastern cooperative oncology group/intergroup trial. J Clin Oncol 2003;21(7):1214–22. https://doi.org/10.1200/JCO.2003.02.005.

[54] Crane CA, Han SJ, Ahn B, et al. Individual patient-specific immunity against high-grade glioma after vaccination with autologous tumor derived peptides bound to the 96 KD chaperone protein. Clin Cancer Res 2013;19(1):205–14. https://doi.org/10.1158/1078-0432.CCR-11-3358.

[55] Bloch O, Crane CA, Fuks Y, et al. Heat-shock protein peptide complex-96 vaccination for recurrent glioblastoma: a phase II, single-arm trial. Neuro-Oncology 2014;16(2):274–9. https://doi.org/10.1093/neuonc/not203.

[56] Wick W, Gorlia T, Bendszus M, et al. Lomustine and bevacizumab in progressive glioblastoma. N Engl J Med 2017;377(20):1954–63. https://doi.org/10.1056/NEJMoa1707358.

[57] Bloch O, Shi Q, Anderson SK, et al. ATIM-14. ALLIANCE A071101: a phase II randomized trial comparing the efficacy of heat shock protein peptide complex-96 (HSPPC-96) vaccine given with bevacizumab versus bevacizumab alone in the treatment of surgically resectable recurrent glioblastoma. Neuro-Oncology 2017;19 (6 suppl).

[58] Bloch O, Lim M, Sughrue ME, et al. Autologous heat shock protein peptide vaccination for newly diagnosed glioblastoma: impact of peripheral PD-L1 expression on response to therapy. Clin Cancer Res 2017;23 (14):3575–84. https://doi.org/10.1158/1078-0432.CCR-16-1369.

[59] Dong H, Strome SE, Salomao DR, et al. Tumor-associated B7-H1 promotes T-cell apoptosis: a potential mechanism of immune evasion. Nat Med 2002;8(8):793–800. https://doi.org/10.1038/nm730.

[60] Berghoff AS, Kiesel B, Widhalm G, et al. Programmed death ligand 1 expression and tumor-infiltrating lymphocytes in glioblastoma. Neuro-Oncology 2015;17(8):1064–75. https://doi.org/10.1093/neuonc/nou307.

CHAPTER

4

Virotherapy treatment of central nervous system tumors

Isaac H. Solomon[a] *and E. Antonio Chiocca*[b]

[a]Department of Pathology, Brigham and Women's Hospital, Harvard Medical School, Boston, MA, United States [b]Department of Neurosurgery, Brigham and Women's Hospital, Harvard Medical School, Boston, MA, United States

Abstract

Virotherapy is a rapidly developing field for the treatment of tumors that are incurable by current standard-of-care therapies, such as glioblastoma in adults and diffuse intrinsic pontine glioma in children. Oncolytic viruses, including both wild-type and genetically modified strains, preferentially infect and kill tumor cells with relative sparing of nonneoplastic cells. Viral immunotherapy acts by stimulating the host immune system to recognize and attack tumor cells. Gene therapy with viral vectors can selectively target neoplastic cells for the activation of prodrugs or expression of cytotoxic or immunomodulatory genes. Many clinical trials have been completed or are currently in progress for adults with newly diagnosed or recurrent high-grade/malignant gliomas as solo agents or in combination with immunotherapeutic agents, including programmed cell death protein 1 inhibitors. This chapter reviews virotherapy clinical trials using adenovirus, herpes simplex virus, measles, parvovirus, poliovirus, reovirus, retroviral replicating vectors (i.e., Toca 511), and vaccinia, with a focus on published results including overall survival and neuropathological characterization. Several trials have shown early promise due to enhanced targeting and death of tumor cells and/or enhanced immune system activation; however, definitive survival benefit has yet to be demonstrated in phase 3 trials, necessitating further study.

Abbreviations

Ad	adenovirus
CD	cluster of differentiation
CDKN2A/B	cyclin-dependent kinase inhibitor 2A/B
CEA	carcinoembryonic antigen
CNS	central nervous system
EGFR	epidermal growth factor receptor
FDA	Food and Drug Administration
GBM	glioblastoma
HCMV	human cytomegalovirus

https://doi.org/10.1016/B978-0-12-819755-4.00005-9

HSV	herpes simplex virus
ICP34.5	infected cell protein 34.5
IDH	isocitrate dehydrogenase
IE	immediate early
IHC	immunohistochemistry
IL-12	interleukin 12
IRES	internal ribosome entry site
MV	measles virus
NK	natural killer
PD-1	programmed cell death protein 1
PD-L1	programmed death ligand 1
PTEN	phosphatase and tensin homolog
PV	poliovirus
TERT	telomerase reverse transcriptase
TK	thymidine kinase
VACV	vaccinia virus
WHO	World Health Organization

Conflict of interest

I.H.S. has no potential conflicts of interest to disclose. E.A.C. is currently an advisor to Advantagene Inc., Alcyone Biosciences, Insightec, Inc., DNAtrix Inc., Immunomic Therapeutics, Sangamo Therapeutics, Seneca Therapeutics, and has equity interest in DNAtrix, Immunomic Therapeutics, and Seneca Therapeutics; he has also advised Oncorus, Merck, Tocagen, Ziopharm, StemGen, NanoTx., Ziopharm Oncology, Cerebral Therapeutics, Genenta, Merck, Janssen, Karcinolys, and Shanghai Biotech. He has received research support from NIH, US Department of Defense, American Brain Tumor Association, National Brain Tumor Society, Alliance for Cancer Gene Therapy, Neurosurgical Research Education Foundation, Advantagene, NewLink Genetics, and Amgen. He is also a named inventor on patents related to oncolytic HSV1 and noncoding RNAs.

Introduction

Despite significant advances in neurosurgical techniques, chemotherapeutic agents, and radiation therapy, malignant brain tumors continue to cause significant morbidity and mortality. In adults, glioblastoma, IDH-wild-type, WHO grade IV, is both the most common and most deadly primary central nervous system tumor, with the median survival of approximately 15 months with standard-of-care therapy, including gross total resection followed by temozolomide and radiation [1]. The infiltrative growth pattern is invariably associated with tumor cell spread beyond grossly involved brain, precluding the type of complete surgical excision which could be achieved for either circumscribed CNS tumors or with the removal of the entire affected organ for extra-CNS sites. While several genomic abnormalities are common in these tumors—typically gain of chromosome 7, focal amplification of EGFR, chromosome 9p loss including homozygous loss of CDKN2A and CDKN2B, loss of chromosome 10 including PTEN, and mutation of TERT promoter gene [2]—numerous other mutations, copy number alterations, and gene rearrangements are recurrently identified. This genetic heterogeneity creates substantial challenges for the adoption of broadly effective treatments, but does allow for the development of targeted therapies for specific pathways which can be identified by gene panels or whole exome/genome sequencing of tumors [3]. In addition to intertumoral differences, intratumoral variations consisting of subclonal populations of cells with more aggressive phenotypes can be present [4]. These clones are often resistant to initial therapies and may become predominant in recurrent/residual tumors. Other high-grade infiltrating gliomas (WHO grade III or IV) in adults, including diffuse midline glioma, H3K27M-mutant, glioblastoma, IDH-mutant, anaplastic astrocytoma, IDH-mutant, anaplastic oligodendroglioma, IDH-mutant and 1p/19q-codeleted, anaplastic pleomorphic xanthroastrocytoma, and anaplastic ependymoma, share similar treatment challenges. High-grade pediatric brain tumors, including medulloblastoma, atypical/teratoid rhabdoid tumor, and diffuse intrinsic pontine glioma, are typically difficult or impossible to completely resect due to midline locations and involvement of brainstem [1]. Combinations of chemotherapy and radiation have some benefit, but also cause significant morbidity due to toxic side effects on the developing brain. Because of these challenges, novel options are needed to treat these aggressive brain tumors in children and adults while minimizing therapeutic side effects.

Virotherapy is a rapidly expanding field for the treatment of a variety of neoplasms, including primary CNS tumors. A wide range of viruses are being studied in animal models and in humans to promote tumor cell death through direct lysis, through the stimulation of an antitumor immune response, or through targeted gene delivery. Utilization of viruses has several advantages over other treatment modalities, including intrinsic selectivity for specific tissue sites and cell types which can be further narrowed through genetic modification. Several viruses are well known to infect the CNS, either as part of a disseminated systemic illness, such as influenza and measles, or as brain or spinal cord-predominant disease, as occurs in arbovirus encephalitides and poliomyelitis, respectively. Most viruses enter the CNS via hematogenous spread, and less commonly enter via retrograde axonal transport (e.g., rabies and herpes simplex virus-1). Viruses may predominately infect a single cell type, such as JC virus and oligodendrocytes, while others such as HSV-1 can recognize multiple cellular receptors and infect virtually any cell type [5]. While some wild-type strains of virus can be used for treatment, therapeutic viruses are often genetically modified to increase tumor specificity and to minimize infection and damage to nonneoplastic tissues and cells. Attenuated viruses must also contend with acquired host immunity due to prior infection or immunization, which may limit both therapeutic and off-target viral infectivity and replication. While mouse tumor models have facilitated the development of many therapeutic viruses, confirmation of safety and efficacy requires well-designed clinical trials in humans. The main focus of this chapter, therefore, is to review viruses that have been tested in humans with primary brain tumors, identified by registration at ClinicalTrials.gov (Table 1). An emphasis has been placed on completed studies with published results (as of October 2019), including survival outcomes and histopathological correlates of virotherapy activity (Fig. 1).

Adenovirus

Adenoviruses are double-stranded DNA viruses in the *Adenoviridae* family that frequently cause mild respiratory infections but can also cause severe disseminated disease involving the central nervous system. Engineered adenoviruses based on human serotype 5 have been utilized as oncolytic viruses as well as vectors for targeted gene expression in a number of therapeutic settings [6].

TABLE 1 Selected CNS tumor virotherapy trials registered at ClinicalTrials.gov (as of October 2019)

Study	Phase	Virus (+/− additional therapy)	Age	Tumor	Trial status	Subjects
NCT00751270	Phase 1b	Ad-HSV-tk + valacyclovir (radiotherapy, chemotherapy)	>18	HGG, newly diagnosed	Completed	15
NCT00589875	Phase 2	Ad-HSV-tk + valacyclovir (radiotherapy, chemotherapy)	>18	HGG, newly diagnosed	Completed	52
NCT00002824	Phase 1	Ad-HSV-tk + ganciclovir	>18	HGG, recurrent	Completed	18
NCT00870181	Phase 2	Ad-HSV-tk + ganciclovir	>18	HGG, recurrent	Completed	47
NCT03603405	Phase 1/2	Ad-HSV-tk + valacyclovir (radiotherapy, chemotherapy)	>18	HGG, newly diagnosed	Recruiting	62
NCT03596086	Phase 1/2	Ad-HSV-tk + valacyclovir (radiotherapy, chemotherapy)	>18	HGG, recurrent	Recruiting	62
NCT03576612	Phase 1	Ad-HSV-tk + valacyclovir (nivolumab, radiotherapy, chemotherapy)	>18	HGG, newly diagnosed	Recruiting	36
NCT01811992	Phase 1	Ad-hCMV-TK & Ad-hCMV-Flt3L + valacyclovir	18–75	HGG, newly diagnosed	Active, not recruiting	19
NCT00031083	Phase 1	Ad BG00001	>18	HGG, recurrent	Suspended	11
NCT02026271	Phase 1	Ad-RTS-hIL-12 + veledimex	18–75	HGG, recurrent	Active, not recruiting	48
NCT03679754	Phase 1	Ad-RTS-hIL-12 + veledimex	18–75	GBM, recurrent	Active, not recruiting	36
NCT03636477	Phase 1	Ad-RTS-hIL-12 + veledimex (nivolumab)	18–75	GBM, recurrent	Recruiting	18
NCT04006119	Phase 2	Ad-RTS-hIL-12 + veledimex (cemiplimab)	18–75	GBM, recurrent	Recruiting	30
NCT03330197	Phase 1	Ad-RTS-hIL-12 + veledimex	<21	Brain tumor, recurrent; DIPG, newly diagnosed	Active, not recruiting	24
NCT00805376	Phase 1	Ad DNX-2401	>18	HGG, recurrent	Completed	37
NCT02197169	Phase 1b	Ad DNX-2401 (interferon-gamma)	>18	GBM, recurrent	Completed	37
NCT01956734	Phase 1	Ad DNX-2401 (temozolomide)	18–75	HGG, recurrent	Completed	31
NCT01582516	Phase 1/2	Ad DNX-2401	18–75	GBM, recurrent	Completed	20
NCT02798406	Phase 2	Ad DNX-2401 (pembrolizumab)	>18	GBM, recurrent	Active, not recruiting	49
NCT03178032	Phase 1	Ad DNX-2401	1–18	DIPG, newly diagnosed	Recruiting	12

TABLE 1 Selected CNS tumor virotherapy trials registered at ClinicalTrials.gov (as of October 2019)—cont'd

Study	Phase	Virus (+/− additional therapy)	Age	Tumor	Trial status	Subjects
NCT03896568	Phase 1	Ad DNX-2401 BM-hMSCs	>18	HGG, recurrent	Recruiting	36
NCT03714334	Phase 1	Ad DNX-2440	>18	GBM, recurrent	Recruiting	24
NCT00028158	Phase 1/2	HSV G207	>19	HGG, recurrent	Completed	65
NCT00157703	Phase 1	HSV G207 (radiation)	>19	HGG, recurrent	Completed	9
NCT02457845	Phase 1	HSV G207 (radiation)	3–18	Any malignant, recurrent	Recruiting	18
NCT03911388	Phase 1	HSV G207 (radiation)	3–18	Any malignant, recurrent	Recruiting	15
NCT02031965	Phase 1	HSV1716	12–21	HGG, recurrent	Terminated	2
NCT03152318	Phase 1	HSV rQNestin34.5v.2 (cyclophosphamide)	>18	HGG, recurrent	Recruiting	108
NCT03657576	Phase 1	HSV C134	>18	HGG, recurrent	Recruiting	24
NCT02062827	Phase 1	HSV M032	>18	HGG, recurrent	Recruiting	36
NCT00390299	Phase 1	MV-CEA	>18	HGG, recurrent	Active, not recruiting	23
NCT02962167	Phase 1	MV-NIS	1–39	MB, AT/RT, recurrent	Recruiting	46
NCT01301430	Phase1/2	Parvovirus H-1	>18	GBM, recurrent	Completed	18
NCT01491893	Phase 1	PVSRIPO	>18	GBM, recurrent	Active, not recruiting	61
NCT02986178	Phase 2	PVSRIPO	>18	HGG, recurrent	Recruiting	102
NCT03043391	Phase 1b	PVSRIPO	12–21	HGG, recurrent	Recruiting	12
NCT03973879	Phase 1/2	PVSRIPO (atezolizumab)	>18	HGG, recurrent	Not yet recruiting	31
NCT00528684	Phase 1	Reovirus	>18	HGG, recurrent	Completed	15
NCT02444546	Phase 1	Reovirus (sargramostim)	10–21	WHO3/4, recurrent; DIPG, newly diagnosed	Active, not recruiting	6
NCT01470794	Phase 1	Toca 511	18–75	HGG, recurrent	Completed	58
NCT02414165	Phase 2/3	Toca 511 (vs lomustine, temozolomide, bevacizumab)	18–75	HGG, recurrent	Active, not recruiting	403
NCT03294486	Phase 1/2	Vaccinia TG6002+5-flucytosine	>18	GBM, recurrent	Recruiting	78

Abbreviations: *Ad*, adenovirus; *HSV*, herpes simplex virus; *TK*, thymidine kinase; *CMV*, cytomegalovirus; *MV*, measles virus; *GBM*, glioblastoma; *HGG*, high-grade glioma; *MB*, medulloblastoma; *AT/RT*, atypical teratoid rhabdoid tumor; *WHO*, World Health Organization.

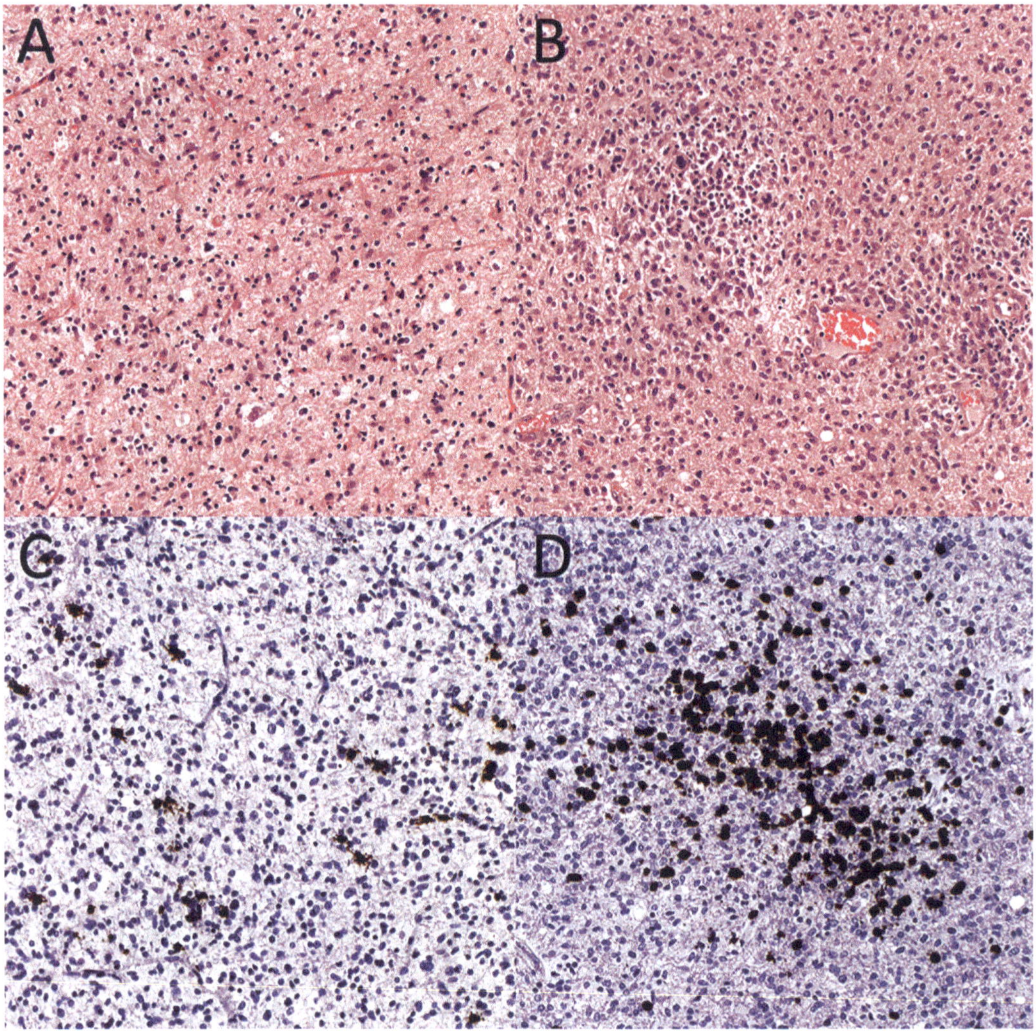

FIG. 1 Histopathologic features of virotherapy. Representative images of recurrent/residual glioblastoma, IDH-wild-type, WHO grade IV in a 55-year-old male pre- (panels A and C) and 50 days posttreatment (panels B and D) with Ad-RTS-hIL-12 with veledimex. Infiltrating high-grade glioma is present in hematoxylin and eosin-stained sections from both time points (panels A and B), with increased cellular density at subsequent recurrence (panel B). A moderate to focally marked increase in inflammatory infiltrates was observed throughout the tumor, including many CD8+ cytotoxic T cells (C and D). Typical features of oncolytic virotherapy not present here include large areas of nonpalisading necrosis in proximity to the injection site(s), and demonstration of viral antigen in necrotic or adjacent cells by immunohistochemistry. All images taken with 20× objective.

Replication-deficient adenoviruses encoding the herpes simplex virus thymidine kinase gene (Ad-HSV-tk) promote the activation of antiviral drugs to cytotoxic forms, selectively killing actively dividing cells [7]. A phase 1 trial in 13 adults with recurrent high-grade glioma reported a median overall survival of 4.0 months with dose-limiting effects at 2×10^{12} vector particles [8]. Postmortem evaluation in six subjects identified viable recurrent tumor with coagulative necrosis and minimal-to-moderate intratumoral lymphohistiocytic infiltrates with minimal extratumoral inflammation. A phase 1b trial in 13 adults with newly diagnosed high-grade glioma was conducted with virus injected into resection cavity followed by oral valacyclovir and standard-of-care therapy, including radiation and temozolomide (NCT00751270) [9]. Median overall survival was 12.4 months with no dose-limiting toxicities. CD3+ T cells were present in all four posttreatment specimens examined, which also revealed CD68+ macrophages. The subsequent phase 2 trial enrolled an additional number of 52 subjects (NCT00589875) [10]. The median overall survival for the phase 1b/2 trial was 17.1 months compared with 13.5 months in standard-of-care-only group ($n = 134$ individuals; $P = .0417$) with no dose-limiting toxicities. Other studies using Ad-HSV-tk viruses showed similar findings (NCT00002824) [11–13], including 19 patients with primary or recurrent malignant gliomas showing an increased median survival from 37.7 weeks (in controls) to 62.4 weeks [14], and 53 patients with recurrent high-grade gliomas and a median overall survival of 45.4 weeks (NCT00870181) [15]. Ongoing phase 1/2 studies in adults with newly diagnosed (NCT03603405; estimated enrollment 62; study completion December 2025; last update August 26, 2019) or recurrent (NCT03596086; estimated enrollment 62; study completion December 2025; last update August 26, 2019) anaplastic astrocytoma or glioblastoma, in combination with standard-of-care therapy, are actively recruiting. Combination therapy of Ad-HSV-TK with anti-PD-1 immune checkpoint inhibitor nivolumab is also being investigated in adults with newly diagnosed high-grade glioma (NCT03576612; estimated enrollment 36; study completion January 2022; last update June 20, 2019). A phase 1 trial with a combination of Ad-hCMK-TK and Ad-hCMV-Fit3L, a cytokine that causes the proliferation of dendritic cells, with valacyclovir for the treatment of adult newly diagnosed high-grade glioma has enrolled 19 participants and has an expected completion date of December 2020 (NCT01811992; active not recruiting; last updated April 11, 2019) [16].

BG00001, a replication-deficient adenovirus expressing interferon beta, has been tested in 11 adults with recurrent high-grade gliomas (NCT00031083) [17]. Virus was intratumorally injected followed by resection at 4 or 8 days with additional injections at the edge of the resection cavity. Resected tumor showed necrosis; infiltration by neutrophils, macrophages, and monocytes; and an increase in apoptosis compared with pretreated tumor. Median overall survival was 17.9 weeks, with dose-limiting toxicity observed in one subject who developed confusion, swelling, and edema, followed by progression, and death 83 days post initial injection. Postmortem examination showed severe necrosis and inflammation involving the leptomeninges and brain surface at the injection site, consistent with direct contact with CSF containing virus.

Ad-RTS-hIL-12 is a replication-deficient adenovirus expressing human interleukin 12 (hIL-12), an interferon gamma-stimulating proinflammatory cytokine with antineoplastic properties, regulated by an orally administered activator ligand veledimex [18]. Results of a phase 1 trial were recently published for 31 adults with recurrent high-grade glioma treated with Ad-RTS-hIL-12 injected in postresection cavity wall followed by 14 days of oral veledimex (NCT02026271) [19]. Median overall survival was 12.7 months. Due to an increased frequency of severe but rapidly reversible adverse events, including cytokine release syndrome, the optimal veledimex dose was determined to be 20 mg. Posttreatment tissue was evaluated in five patients at 1–6 months post injection, which showed recurrent/residual tumor with increased inflammatory infiltrates, including CD8+ T cells, compared with pretreatment biopsies. Multiplexed immunofluorescence staining of three paired pre- and posttreatment samples showed increases in CD3-, CD8-, PD-1-, and PD-L1-expressing cells, with no differences in FoxP3+ T cells or CD56+ NK cells. An expansion study of 36 additional adults with recurrent glioblastoma is currently active but no longer recruiting (NCT03679754; estimated completion June 2021; last updated February 15, 2019). Additional studies currently recruiting participants include combination therapy with anti-PD-1 monoclonal antibodies, which function as immune checkpoint inhibitors. A phase 1 trial with nivolumab is recruiting 18 adults with recurrent glioblastoma (NCT03636477; estimated completion December 2021; last updated March 22, 2019) and a phase 2 trial with cemiplimab-rwlc is recruiting 30 participants (NCT04006119; estimated completion August 2022; last updated August 15, 2019). In addition, a phase 1 study is actively recruiting 30 pediatric patients with recurrent/progressive supratentorial tumors or diffuse intrinsic pontine glioma (NCT03330197; estimated completion March 2021; last updated June 17, 2019).

As an alternative to gene therapy, oncolytic adenoviruses have also been studied for utility in treating high-grade gliomas [20]. DNX-2401 (Delta-24-RGD), which kills cells by direct oncolytic action, has increased tumor selectivity through the deletion of 24 base pairs in E1A gene preventing replication in retinoblastoma gene pathway functional cells combined with the insertion of RGD-motif into fiber H-loop to increase cell uptake by $\alpha_v\beta_3/\beta_5$ integrins. The results of a phase 1 trial in 37 adults with recurrent malignant gliomas have recently been reported (NCT00805376) [21]. Patients either received a single intratumoral injection of virus following biopsy or received catheter-administered virus followed by en bloc resection 14 days later with additional virus injection in resection cavity. No dose-limiting toxicity or severe adverse events were encountered. Median overall survival time for the dose escalation arm was 9.5 months, with 5/25 (20%) of patients surviving greater than 3 years, and 13.0 months for resection and reinjection arms. Pathologic evaluation of day 14 posttreatment samples demonstrated viral antigens (E1A or hexon) in 6/11 (55%) of tumors, associated with viral inclusions, macrophages and CD8+ T cells, and necrosis. Quantification of immune cells showed increases in CD4 and CD8 cell density compared with pretreatment specimens, with an increase in checkpoint protein TIM-3 and no change in PD-1, or PD-L1, IDO-1H. Tumor resected at 1.5 months posttreatment was predominantly necrotic, with acute and chronic inflammation, but no adenovirus-infected cells. A subsequent trial of DNX-2401 was conducted in adults with recurrent glioblastoma with subcutaneous interferon gamma, which was poorly tolerated and showed no improvement in survival compared with DNX-2401 alone (NCT02197169) [22]. Combination treatment with temozolomide in 31 adults with recurrent glioma has been well tolerated with no severe toxicities and survival of 3 patients at least 19 months posttreatment (NCT01956734; last updated October 24, 2017) [23]. A phase 1/2 in adults with recurrent glioblastoma was completed in December 2014 but has not yet published results (NCT01582516; last updated March 9, 2015). Ongoing studies include an active, but not recruiting phase 2 trial of DNX-2401 in combination with pembrolizumab in 49 adults with recurrent glioblastoma (NCT02798406; estimated completion June 2021; last updated March 22, 2019) and a phase 1 trial currently recruiting 12 patients with newly diagnosed diffuse intrinsic pontine gliomas (NCT03178032; estimated completion December 2019, last updated August 7, 2019) [24]. A phase 1 trial with DNX-2401 delivered by intraarterial injection of infected mesenchymal stem cells is recruiting 36 adults with recurrent high-grade gliomas (NCT03896568;

estimated completion February 2020; last updated April 1, 2019). A related virus DNX-2440, engineered to express OX40 ligand, a costimulatory immune checkpoint factor, is in a phase 1 study currently recruiting 24 adults with recurrent glioblastoma (NCT03714334; estimated completion October 2022; last updated November 9, 2018) [25].

Herpes simplex virus

Herpes simplex virus 1 (HSV-1) is a double-stranded DNA virus in the *Herpesviridae* family that infects a variety of tissues and cell types, and can cause a severe, often fatal encephalitis. Several genetically modified oncolytic herpes viruses have been studied for the treatment of brain tumors in humans [26].

G207 was created through the modification of HSV-1 laboratory F strain with the deletion of infected cell protein 34.5 (ICP34.5) encoded by $\gamma_1 34.5$, necessary for replication in nonneoplastic cells, and lacZ insertion inactivating the ICP6 gene ($U_L 39$), encoding the large subunit of the viral ribonucleotide reductase. In a phase 1 study with 21 adults with recurrent high-grade astrocytic tumors, no serious toxicities or evidence of encephalitis was identified [27]. Six patients underwent posttreatment biopsy, and five patients underwent autopsy evaluation. Two biopsies were positive for virus by PCR, while none of the tissues showed evidence of encephalitis or HSV antigens by immunohistochemistry. A subsequent study involving multiple inoculations in tumor and postresection cavity in six adults showed increased CD3-, CD8-, and HAM56-positive cells with the detection of virus by PCR and IHC targeting late protein glycoprotein C (NCT00028158) [28]. A combination of G207 with a single 5Gy radiation dose was investigated in nine adults with recurrent malignant glioma (NCT00157703) [29]. This study showed stable disease or partial response in 6/9 (67%) patients for at least one time point, and median survival of 7.5months from injection. Stereotactic biopsy in one patient 2months posttreatment showed progression from anaplastic astrocytoma to glioblastoma, minimal inflammation, no evidence of encephalitis, and few HSV-positive cells by immunohistochemistry. Autopsy at 8.2months did not show inflammation, viral cytopathic effects, or chronic demyelination. Trials are currently underway in recurrent or refractory pediatric supratentorial (NCT02457845; estimated completion October 2020; last updated September 13, 2019) and cerebellar (NCT03911388; estimated completion September 2023; last updated September 17, 2019) brain tumors with and without radiation [30, 31].

Multiple studies have been conducted with HSV1716, created from Glasgow strain 17 by the deletion of both copies of the ICP34.5 gene allowing the virus to replicate in actively dividing but not terminally differentiated cells. In an initial study of nine adults with high-grade gliomas, intratumoral injection did not cause encephalitis or adverse symptoms [32]. Posttreatment surgical and autopsy tissues collected 3.5weeks to 6months after injection in five patients showed high-grade glioma without immunohistochemical or PCR evidence of HSV. A subsequent study in 12 patients was conducted with the resection of tumor 4–9days following intratumoral injection of HSV1716 [33]. Virus was detected by culture in 2/12 (8%) patients, PCR in 10/12 (83%) tumors at injection site, and 4/10 (40%) from distal tumor sites, and by immunohistochemistry in two patients, with no evidence of toxicity or encephalitis. In a third trial with HSV1716, virus was injected into the resection cavity rim in 12 adult patients with newly diagnosed or recurrent high-grade gliomas, with no virus-associated toxicity observed [34]. Cultured tumor cells from four patients collected at the time of virus injection showed an efficient replication of virus in vitro. A clinical trial in pediatric high-grade gliomas was terminated after enrollment of two patients (NCT02031965; last updated May 30, 2016).

To improve on first-generation oncolytic herpes viruses, additional modifications have been made to improve the activity of ICP34.5-deleted viruses. rQNestin34.5v.2 is an ICP34.5- and U_L39-deleted virus with re-insertion of a single copy of ICP34.5 under control of the nestin promoter, which restricts viral replication to glioma cells compared with nonneoplastic glial and neuronal cells due to the lack of nestin expression [35]. An ongoing phase 1 clinical trial in adults with recurrent high-grade gliomas is recruiting an estimated number of 108 participants for intratumoral administration of virus (NCT03152318; estimated completion July 2022; last updated August 13, 2019). C134 is a chimeric virus in which the human cytomegalovirus (HCMV) protein kinase R evasion gene IRS1 has been added under the control of HCMV IE promoter to increase viral replication without causing neurovirulence associated with ICP34.5 expression [36, 37]. A phase 1 trial in adults with recurrent glioblastoma is recruiting 24 participants for intratumoral administration of virus (NCT03657576; estimated completion September 2024; last updated October 3, 2019). M032 is an ICP34.5-deleted virus that expresses human IL-12 p35 and p40 subunits, which can stimulate an antitumor immune response [38]. A phase 1 trial in adults with recurrent glioblastoma is currently recruiting for intratumoral administration of virus in an estimated number of 36 participants (NCT02062827; estimated completion September 2020; last updated July 13, 2019) [39].

Measles

Measles morbillivirus is a single-stranded RNA virus in the *Paramyxoviridae* family that causes measles (rubeola) and subacute sclerosing panencephalitis. Oncolytic viruses have been developed through the modification of the live-attenuated Edmonston vaccine strain (MV-Edm) [40]. MV-Edm contains a mutation in the viral hemagglutinin gene leading to the utilization of CD46 as a receptor that is frequently overexpressed in tumors, as well as CD150 (signaling lymphocyte activation molecular) and nectin-4 (polio virus receptor-related 4). MV-CEA has been modified from MV-Edm to encode the gene for the soluble extracellular N-terminal domain of human carcinoembryonic antigen (CEA), which is secreted in serum and can be used to track viral gene expression. A phase 1 trial involving 23 adults with recurrent high-grade gliomas is currently active, not recruiting (NCT00390299; estimated completion November 2019; last updated January 4, 2019). Subjects either underwent immediate resection of tumor with the injection of virus into the resection cavity, or intratumoral injection followed by resection 5 days later with injection into resection cavity. Pending histological evaluation of the resected tissue will allow for evaluation of the viral activity. No dose-limiting toxicities were reported to have been observed using intracranial doses up to 2×10^7 $TCID_{50}$ [41]. A separate oncolytic virus, MV-NIS, has been modified from MV-EdM to encode the gene for sodium iodide symporter (NIS), which facilitates the intracellular concentration of radioisotopes which can be used to track viral spread via positron emission tomography or single-photon emission computed tomography. An ongoing clinical trial in children and young adults with recurrent medulloblastoma and atypical teratoid/rhabdoid tumor is currently recruiting (NCT02962167; estimated completion February 2022; last update posted August 21, 2019). Forty-six subjects will be injected with virus into tumor bed following resection for locally recurrent disease or injected into cerebrospinal fluid via lumbar puncture for disseminated recurrent disease and followed for 2 years.

Parvovirus

Parvoviruses are single-stranded DNA viruses in the *Parvoviridae* family. Parvovirus H-1 (H-1PV; ParvOryx) naturally infects rats and is generally considered to be nonpathogenic in humans [42]. H-1PV has been shown to have oncolytic properties, including activity against glioma cell lines and rat glioma models through the activation of cathepsins to induce apoptosis [43]. A phase 1 study has been completed in adults with recurrent glioblastoma, in which one group received intratumorally injected virus and the second group received virus intravenously, both of which were followed by resection and injection of virus around the resection cavity (NCT01301430) [44]. In 18 treated patients, no dose-limiting toxicity or maximum tolerated dose were identified. During the 6-month follow-up period, 12/18 (67%) subjects had recurrent disease or died. Median overall survival was 464 days. Virus was confirmed to have crossed the blood-brain barrier, and new antibodies were detected in subjects with higher doses of virus. Posttreatment resection tissue was collected for all subjects, and viral DNA was positive in 11/12 (92%) of intratumorally treated and 3/6 (50%) of intravenously treated patients by fluorescence in situ hybridization. The presence of virus was confirmed by RT-PCR, and sequencing of viral genome did not identify any new mutations, indicating stability of the virus in vivo. Prominent tumor-infiltrating leukocytes were observed, consisting predominantly of CD8 cytotoxic T cells and CD4-positive cells, although no pretreatment tissue was available for quantitative comparison.

Poliovirus

Polioviruses are single-stranded RNA viruses in the *Enterovirus* genus and *Picornaviridae* family responsible for poliomyelitis. PVSRIPO (PV (Sabin)-Rhinovirus IRES PV Open reading frame) is an oncolytic virus currently under investigation in malignant brain tumors, derived from the live attenuated Sabin type 1 poliovirus vaccine in which the endogenous internal ribosomal entry site (IRES) has been replaced with the sequence from the human rhinovirus type 2 [45]. A neuro-attenuated phenotype results, retaining the cytotoxic and interferon response activation in glioblastoma cells which highly express the poliovirus receptor, CD155, a member of the nectin family of cell adhesion molecules [46]. The results of a phase 1 trial in 61 adults with recurrent glioblastoma intratumorally injected have recently been reported (NCT01491893) [47]. The median overall survival was 12.5 months compared with 11.3 months in the historical control group, which differed from the treated group by lower extent of resection, higher previous treatment failure with bevacizumab, and unknown IDH1/2 status. The excitement generated by this trial though had to do with 8/61 subjects with a durable response for at least 2 years. Treatment was generally well tolerated with grade 3 or above adverse events in 10/52 (19%) of subjects in the dose-expansion phase, with no evidence of encephalitis, myelitis, or meningitis. Four patients were reported to have undergone reoperation, and brain tissue was collected from 14 patients at autopsy, which all showed the presence of glioblastoma. A phase 2 trial in adults (NCT02986178) and a phase 1 trial in children (NCT03043391) with recurrent high-grade gliomas are currently recruiting, and a phase 1/2 trial in combination with atezolizumab (antiprogrammed cell death-ligand 1 antibody) (NCT03973879) is expected to start recruiting in October 2019 (last updated August 20, 2019).

Reovirus

While reoviruses, double-stranded RNA viruses in the *Reoviridae* family, are generally considered nonpathogenic in humans, oncolytic properties have been observed in a broad range of tumor types, including carcinoma, myeloma, and melanoma, and the wild-type strain pelareorep (Reolysin) has received orphan drug designation by the FDA [48]. Activation of Ras signaling pathways in tumor cells inhibits protein kinase R activity, which normally functions in the detection of double-stranded viral RNA, causing the inhibition of protein synthesis and promotion of apoptosis [49]. A phase 1 dose escalation trial has been conducted in adults with recurrent glioblastoma (NCT00528684) [50]. In 15 subjects with intratumoral infusion of virus, no dose-limiting toxicity or maximum tolerated dose was identified. During 24 weeks of the study period, 10 patients had stable disease, one had a partial response, and four had progressive disease, with an overall median survival time of 140 days. No posttreatment tumor samples were available for analysis. An active but not recruiting study (NCT02444546; estimated completion April 2020; last updated July 24, 2019) is in progress for recurrent high-grade pediatric brain tumors, including malignant gliomas, medulloblastomas, and atypical teratoid/rhabdoid tumor, as well as diffuse intrinsic pontine gliomas. Six patients have been enrolled, and will receive 12 monthly courses of reovirus intravenously and sargramostim (GM-CSF) subcutaneously, followed by observation for up to 5 years.

Toca 511

Retroviral replicating vectors, single-stranded RNA viruses in the *Retroviridae* family, including Toca 511 (vocimagene amiretrorepvec), are based on murine leukemia virus and do not directly cause cell lysis or stimulate an immune response [51]. Instead, virus is intracranially or intravenously administered, and preferentially infects and replicates in malignant cells due to the increased proliferative rate and lack of immune response. The virus carries the gene for yeast cytosine deaminase, which converts orally administered 5-fluorocytosine to 5-fluorouracil, and causes cytotoxicity of infected and neighboring cells. A phase 1 trial was reported for 56 adults with recurrent high-grade gliomas (NCT01470794) [52]. Toca 511 was injected into resection cavity beds, followed by weekly Toca FC (extended-release 5-fluorocytosine) cycles every 6 weeks. Median survival was 11.9 months, with complete radiological responses observed in 11% of the subjects, and no secondary malignancies or lymphoproliferative disorders have been detected. Posttreatment tissue was not available for evaluation, but flow cytometry of peripheral blood identified treatment-associated increase in proliferating T cells in subjects without progression compared with subjects with progression. Based on these results, a phase 2/3 trial with 403 adults with glioblastoma or anaplastic astrocytoma comparing Toca 511 + Toca FC with standard-of-care therapy (NCT02414165; active, not recruiting; last updated November 30, 2018; estimated completion date March 2023) has been conducted. A press release by Tocagen Inc. on 9/12/19 reported a decreased survival compared with standard of care (11.1 months median compared to 12.2 months, HR = 1.06, $P = .6154$) [53]. Phase 2/3 studies in newly diagnosed high-grade gliomas are currently planned but have not yet started recruiting (NRG-BN006).

Vaccinia virus

Wild-type and recombinant vaccinia viruses (VACV), double-stranded DNA viruses in the *Poxviridae* family, have been tested preclinically and clinically in a variety of solid tumors with promising results [54]. TG6002 is derived from the Copenhagen strain of VACV, originally used as a smallpox vaccine in Denmark and the Netherlands, which contains strong oncolytic viral activity [55]. TG6002 combines the deletion of the gene encoding the large subunit of viral ribonucleotide reductase, *I4L*, and the replacement of the viral thymidine kinase-encoding gene *J2R* with *FCU1*, a gene encoding a fusion protein with cytosine deaminase and uracil phosphoribosyltransferase activities. The *I4L* and *J2R* deletions promote viral replication selectively in tumor cells, which tend to contain elevated levels of thymidine triphosphate and deoxynucleotide diphosphate, substrates required for viral DNA synthesis and replication. *FCU1* expression catalyzes the conversion of orally administered 5-fluorocytosine into 5-fluorouracil and 5-fluorouracil-monophosphate, which causes cytotoxicity through the inhibition of DNA and protein synthesis. Combined with VACV oncolytic properties, these modifications are hypothesized to increase tumor lysis and decrease toxicity of nonneoplastic cells. A phase 1/2 trial in 78 adults with progression of glioblastoma after standard-of-care therapy is currently in progress (NCT03294486; estimated completion September 2021; last updated November 6, 2017).

Conclusion

Virotherapy is a rapidly developing field in the treatment of primary CNS tumors. The phase 1/2 clinical trials reviewed above have shown a significant promise for improved patient outcomes that have yet to be definitively demonstrated in phase 3 trials. Many studies have been conducted in patients with recurrent tumors, making overall survival difficult to compare due to differences in preceding medical care. Significant tumor heterogeneity exists, particularly when trial design lags behind updates to WHO tumor classification, and with the identification of prognostic biomarkers, such as O^6-methylguanine-DNA methyltransferase promoter methylation. Both of these factors complicate the interpretation of survival benefit due to known differences in natural history or response to treatment, making comparisons with historical controls unreliable. Smaller studies without randomized control arms are particularly prone to overestimation of efficacy when occasional longer-term survivors are identified. Histopathological evaluation of pre- and posttreatment tissue has been invaluable in confirming desired tropism and activity of viruses in vivo, including the demonstration of tumor cell lysis and increased immune cell infiltration and activation. In addition to standard histopathology, the generation of patient-derived tumor models, including cell lines and mouse xenografts, robust genomic characterization, and in-depth immunophenotyping are all essential tools for the characterization of clinical trial tissue specimens, and for the development of future viral therapies.

Acknowledgment

E.A.C. and I.H.S. are supported by P01 CA163205.

References

[1] Louis DN, Ohgaki H, Wiestler OD, Cavenee WK. WHO classification of tumours of the central nervous system. Revised 4th ed. Lyon: World Health Organization; 2016.

[2] Brat DJ, Aldape K, Colman H, Holland EC, Louis DN, Jenkins RB, Kleinschmidt-DeMasters BK, Perry A, Reifenberger G, Stupp R, von Deimling A, Weller M. cIMPACT-NOW update 3: recommended diagnostic criteria for "diffuse astrocytic glioma, IDH-wildtype, with molecular features of glioblastoma, WHO grade IV". Acta Neuropathol 2018;136(5):805–10.

[3] Prados MD, Byron SA, Tran NL, Phillips JJ, Molinaro AM, Ligon KL, Wen PY, Kuhn JG, Mellinghoff IK, de Groot JF, Colman H, Cloughesy TF, Chang SM, Ryken TC, Tembe WD, Kiefer JA, Berens ME, Craig DW, Carpten JD, Trent JM. Toward precision medicine in glioblastoma: the promise and the challenges. Neuro-Oncology 2015;17 (8):1051–63.

[4] Kim H, Zheng S, Amini SS, Virk SM, Mikkelsen T, Brat DJ, Grimsby J, Sougnez C, Muller F, Hu J, Sloan AE, Cohen ML, Van Meir EG, Scarpace L, Laird PW, Weinstein JN, Lander ES, Gabriel S, Getz G, Meyerson M, Chin L, Barnholtz-Sloan JS, Verhaak RG. Whole-genome and multisector exome sequencing of primary and post-treatment glioblastoma reveals patterns of tumor evolution. Genome Res 2015;25(3):316–27.

[5] Karasneh GA, Shukla D. Herpes simplex virus infects most cell types in vitro: clues to its success. Virol J 2011;8:481.

[6] Panek WK, Kane JR, Young JS, Rashidi A, Kim JW, Kanojia D, Lesniak MS. Hitting the nail on the head: combining oncolytic adenovirus-mediated virotherapy and immunomodulation for the treatment of glioma. Oncotarget 2017;8(51):89391–405.

[7] Maatta AM, Samaranayake H, Pikkarainen J, Wirth T, Yla-Herttuala S. Adenovirus mediated herpes simplex virus-thymidine kinase/ganciclovir gene therapy for resectable malignant glioma. Curr Gene Ther 2009;9 (5):356–67.

[8] Trask TW, Trask RP, Aguilar-Cordova E, Shine HD, Wyde PR, Goodman JC, Hamilton WJ, Rojas-Martinez A, Chen SH, Woo SL, Grossman RG. Phase I study of adenoviral delivery of the HSV-tk gene and ganciclovir administration in patients with current malignant brain tumors. Mol Ther 2000;1(2):195–203.

[9] Chiocca EA, Aguilar LK, Bell SD, Kaur B, Hardcastle J, Cavaliere R, McGregor J, Lo S, Ray-Chaudhuri A, Chakravarti A, Grecula J, Newton H, Harris KS, Grossman RG, Trask TW, Baskin DS, Monterroso C, Manzanera AG, Aguilar-Cordova E, New PZ. Phase IB study of gene-mediated cytotoxic immunotherapy adjuvant to up-front surgery and intensive timing radiation for malignant glioma. J Clin Oncol 2011;29(27):3611–9.

[10] Wheeler LA, Manzanera AG, Bell SD, Cavaliere R, McGregor JM, Grecula JC, Newton HB, Lo SS, Badie B, Portnow J, Teh BS, Trask TW, Baskin DS, New PZ, Aguilar LK, Aguilar-Cordova E, Chiocca EA. Phase II multicenter study of gene-mediated cytotoxic immunotherapy as adjuvant to surgical resection for newly diagnosed malignant glioma. Neuro-Oncology 2016;18(8):1137–45.

[11] Eck SL, Alavi JB, Alavi A, Davis A, Hackney D, Judy K, Mollman J, Phillips PC, Wheeldon EB, Wilson JM. Treatment of advanced CNS malignancies with the recombinant adenovirus H5.010RSVTK: a phase I trial. Hum Gene Ther 1996;7(12):1465–82.

[12] Smitt PS, Driesse M, Wolbers J, Kros M, Avezaat C. Treatment of relapsed malignant glioma with an adenoviral vector containing the herpes simplex thymidine kinase gene followed by ganciclovir. Mol Ther 2003;7(6):851–8.

[13] Germano IM, Fable J, Gultekin SH, Silvers A. Adenovirus/herpes simplex-thymidine kinase/ganciclovir complex: preliminary results of a phase I trial in patients with recurrent malignant gliomas. J Neuro-Oncol 2003;65 (3):279–89.

[14] Immonen A, Vapalahti M, Tyynela K, Hurskainen H, Sandmair A, Vanninen R, Langford G, Murray N, Yla-Herttuala S. AdvHSV-tk gene therapy with intravenous ganciclovir improves survival in human malignant glioma: a randomised, controlled study. Mol Ther 2004;10(5):967–72.

[15] Ji N, Weng D, Liu C, Gu Z, Chen S, Guo Y, Fan Z, Wang X, Chen J, Zhao Y, Zhou J, Wang J, Ma D, Li N. Adenovirus-mediated delivery of herpes simplex virus thymidine kinase administration improves outcome of recurrent high-grade glioma. Oncotarget 2016;7(4):4369–78.

[16] Puntel M, Ghulam Muhammad AKM, Farrokhi C, Vanderveen N, Paran C, Appelhans A, Kroeger KM, Salem A, Lacayo L, Pechnick RN, Kelson KR, Kaur S, Kennedy S, Palmer D, Ng P, Liu C, Krasinkiewicz J, Lowenstein PR, Castro MG. Safety profile, efficacy, and biodistribution of a bicistronic high-capacity adenovirus vector encoding a combined immunostimulation and cytotoxic gene therapy as a prelude to a phase I clinical trial for glioblastoma. Toxicol Appl Pharmacol 2013;268(3):318–30.

[17] Chiocca EA, Smith KM, McKinney B, Palmer CA, Rosenfeld S, Lillehei K, Hamilton A, DeMasters BK, Judy K, Kirn D. A phase I trial of Ad.hIFN-beta gene therapy for glioma. Mol Ther 2008;16(3):618–26.

[18] Barrett JA, Cai H, Miao J, Khare PD, Gonzalez P, Dalsing-Hernandez J, Sharma G, Chan T, Cooper LJN, Lebel F. Regulated intratumoral expression of IL-12 using a RheoSwitch Therapeutic System((R)) (RTS((R))) gene switch as gene therapy for the treatment of glioma. Cancer Gene Ther 2018;25(5–6):106–16.

[19] Chiocca EA, Yu JS, Lukas RV, Solomon IH, Ligon KL, Nakashima H, Triggs DA, Reardon DA, Wen P, Stopa BM, Naik A, Rudnick J, Hu JL, Kumthekar P, Yamini B, Buck JY, Demars N, Barrett JA, Gelb AB, Zhou J, Lebel F, Cooper LJN. Regulatable interleukin-12 gene therapy in patients with recurrent high-grade glioma: results of a phase 1 trial. Sci Transl Med 2019;11(505).

[20] Jiang H, Gomez-Manzano C, Rivera-Molina Y, Lang FF, Conrad CA, Fueyo J. Oncolytic adenovirus research evolution: from cell-cycle checkpoints to immune checkpoints. Curr Opin Virol 2015;13:33–9.

[21] Lang FF, Conrad C, Gomez-Manzano C, Yung WKA, Sawaya R, Weinberg JS, Prabhu SS, Rao G, Fuller GN, Aldape KD, Gumin J, Vence LM, Wistuba I, Rodriguez-Canales J, Villalobos PA, Dirven CMF, Tejada S, Valle RD, Alonso MM, Ewald B, Peterkin JJ, Tufaro F, Fueyo J. Phase I study of DNX-2401 (Delta-24-RGD) oncolytic adenovirus: replication and immunotherapeutic effects in recurrent malignant glioma. J Clin Oncol 2018;36 (14):1419–27.

[22] Lang FF, Tran ND, Puduvalli VK, Elder JB, Fink KL, Conrad CA, Yung WKA, Penas-Prado M, Gomez-Manzano C, Peterkin J, Fueyo J. Phase 1b open-label randomized study of the oncolytic adenovirus DNX-2401 administered with or without interferon gamma for recurrent glioblastoma. J Clin Oncol 2017;35(15_suppl).

[23] Alonso MM, García-Moure M, Gonzalez-Huarriz M, Marigil M, Hernandez-Alcoceba R, Buñales M, Hervás S, Gallego J, Gomez-Manzano C, Fueyo J, Lang F, Peterkin J, Diez-Valle R, Tejada S. Abstract CT027: oncolytic virus DNX-2401 with a short course of temozolomide for glioblastoma at first recurrence: clinical data and prognostic biomarkers. Cancer Res 2017;77(13 Supplement).

[24] Tejada S, Diez-Valle R, Dominguez PD, Patino-Garcia A, Gonzalez-Huarriz M, Fueyo J, Gomez-Manzano C, Idoate MA, Peterkin J, Alonso MM. DNX-2401, an oncolytic virus, for the treatment of newly diagnosed diffuse intrinsic pontine gliomas: a case report. Front Oncol 2018;8:61.

[25] Jiang H, Rivera-Molina Y, Gomez-Manzano C, Clise-Dwyer K, Bover L, Vence LM, Yuan Y, Lang FF, Toniatti C, Hossain MB, Fueyo J. Oncolytic adenovirus and tumor-targeting immune modulatory therapy improve autologous cancer vaccination. Cancer Res 2017;77(14):3894–907.

[26] Totsch SK, Schlappi C, Kang KD, Ishizuka AS, Lynn GM, Fox B, Beierle EA, Whitley RJ, Markert JM, Gillespie GY, Bernstock JD, Friedman GK. Oncolytic herpes simplex virus immunotherapy for brain tumors: current pitfalls and emerging strategies to overcome therapeutic resistance. Oncogene 2019;38(34):6159–71.

[27] Markert JM, Medlock MD, Rabkin SD, Gillespie GY, Todo T, Hunter WD, Palmer CA, Feigenbaum F, Tornatore C, Tufaro F, Martuza RL. Conditionally replicating herpes simplex virus mutant, G207 for the treatment of malignant glioma: results of a phase I trial. Gene Ther 2000;7(10):867–74.

[28] Markert JM, Liechty PG, Wang W, Gaston S, Braz E, Karrasch M, Nabors LB, Markiewicz M, Lakeman AD, Palmer CA, Parker JN, Whitley RJ, Gillespie GY. Phase Ib trial of mutant herpes simplex virus G207 inoculated pre-and post-tumor resection for recurrent GBM. Mol Ther 2009;17(1):199–207.

[29] Markert JM, Razdan SN, Kuo HC, Cantor A, Knoll A, Karrasch M, Nabors LB, Markiewicz M, Agee BS, Coleman JM, Lakeman AD, Palmer CA, Parker JN, Whitley RJ, Weichselbaum RR, Fiveash JB, Gillespie GY. A phase 1 trial of oncolytic HSV-1, G207, given in combination with radiation for recurrent GBM demonstrates safety and radiographic responses. Mol Ther 2014;22(5):1048–55.

[30] Waters AM, Johnston JM, Reddy AT, Fiveash J, Madan-Swain A, Kachurak K, Bag AK, Gillespie GY, Markert JM, Friedman GK. Rationale and design of a phase 1 clinical trial to evaluate HSV G207 alone or with a single radiation dose in children with progressive or recurrent malignant supratentorial brain tumors. Hum Gene Ther Clin Dev 2017;28(1):7–16.
[31] Bernstock JD, Wright Z, Bag AK, Gessler F, Gillespie GY, Markert JM, Friedman GK, Johnston JM. Stereotactic placement of intratumoral catheters for continuous infusion delivery of herpes simplex virus -1 G207 in pediatric malignant supratentorial brain tumors. World Neurosurg 2019;122:e1592–8.
[32] Rampling R, Cruickshank G, Papanastassiou V, Nicoll J, Hadley D, Brennan D, Petty R, MacLean A, Harland J, McKie E, Mabbs R, Brown M. Toxicity evaluation of replication-competent herpes simplex virus (ICP 34.5 null mutant 1716) in patients with recurrent malignant glioma. Gene Ther 2000;7(10):859–66.
[33] Papanastassiou V, Rampling R, Fraser M, Petty R, Hadley D, Nicoll J, Harland J, Mabbs R, Brown M. The potential for efficacy of the modified (ICP 34.5(-)) herpes simplex virus HSV1716 following intratumoural injection into human malignant glioma: a proof of principle study. Gene Ther 2002;9(6):398–406.
[34] Harrow S, Papanastassiou V, Harland J, Mabbs R, Petty R, Fraser M, Hadley D, Patterson J, Brown SM, Rampling R. HSV1716 injection into the brain adjacent to tumour following surgical resection of high-grade glioma: safety data and long-term survival. Gene Ther 2004;11(22):1648–58.
[35] Kambara H, Okano H, Chiocca EA, Saeki Y. An oncolytic HSV-1 mutant expressing ICP34.5 under control of a nestin promoter increases survival of animals even when symptomatic from a brain tumor. Cancer Res 2005;65 (7):2832–9.
[36] Shah AC, Parker JN, Gillespie GY, Lakeman FD, Meleth S, Markert JM, Cassady KA. Enhanced antiglioma activity of chimeric HCMV/HSV-1 oncolytic viruses. Gene Ther 2007;14(13):1045–54.
[37] Ghonime MG, Jackson J, Shah A, Roth J, Li M, Saunders U, Coleman J, Gillespie GY, Markert JM, Cassady KA. Chimeric HCMV/HSV-1 and Deltagamma134.5 oncolytic herpes simplex virus elicit immune mediated antigliomal effect and antitumor memory. Transl Oncol 2018;11(1):86–93.
[38] Roth JC, Cassady KA, Cody JJ, Parker JN, Price KH, Coleman JM, Peggins JO, Noker PE, Powers NW, Grimes SD, Carroll SL, Gillespie GY, Whitley RJ, Markert JM. Evaluation of the safety and biodistribution of M032, an attenuated herpes simplex virus type 1 expressing hIL-12, after intracerebral administration to aotus nonhuman primates. Hum Gene Ther Clin Dev 2014;25(1):16–27.
[39] Patel DM, Foreman PM, Nabors LB, Riley KO, Gillespie GY, Markert JM. Design of a phase I clinical trial to evaluate M032, a genetically engineered HSV-1 expressing IL-12, in patients with recurrent/progressive glioblastoma multiforme, anaplastic astrocytoma, or gliosarcoma. Hum Gene Ther Clin Dev 2016;27(2):69–78.
[40] Aref S, Bailey K, Fielding A. Measles to the rescue: a review of oncolytic measles virus. Viruses 2016;8(10).
[41] Msaouel P, Opyrchal M, Dispenzieri A, Peng KW, Federspiel MJ, Russell SJ, Galanis E. Clinical trials with oncolytic measles virus: current status and future prospects. Curr Cancer Drug Targets 2018;18(2):177–87.
[42] Marchini A, Bonifati S, Scott EM, Angelova AL, Rommelaere J. Oncolytic parvoviruses: from basic virology to clinical applications. Virol J 2015;12:6.
[43] Di Piazza M, Mader C, Geletneky K, Herrero YCM, Weber E, Schlehofer J, Deleu L, Rommelaere J. Cytosolic activation of cathepsins mediates parvovirus H-1-induced killing of cisplatin and TRAIL-resistant glioma cells. J Virol 2007;81(8):4186–98.
[44] Geletneky K, Hajda J, Angelova AL, Leuchs B, Capper D, Bartsch AJ, Neumann JO, Schoning T, Husing J, Beelte B, Kiprianova I, Roscher M, Bhat R, von Deimling A, Bruck W, Just A, Frehtman V, Lobhard S, Terletskaia-Ladwig E, Fry J, Jochims K, Daniel V, Krebs O, Dahm M, Huber B, Unterberg A, Rommelaere J. Oncolytic H-1 parvovirus shows safety and signs of immunogenic activity in a first phase I/IIa glioblastoma trial. Mol Ther 2017;25(12):2620–34.
[45] Gromeier M, Nair SK. Recombinant poliovirus for cancer immunotherapy. Annu Rev Med 2018;69:289–99.
[46] Chandramohan V, Bryant JD, Piao H, Keir ST, Lipp ES, Lefaivre M, Perkinson K, Bigner DD, Gromeier M, McLendon RE. Validation of an immunohistochemistry assay for detection of CD155, the poliovirus receptor, in malignant gliomas. Arch Pathol Lab Med 2017;141(12):1697–704.
[47] Desjardins A, Gromeier M, Herndon 2nd JE, Beaubier N, Bolognesi DP, Friedman AH, Friedman HS, McSherry F, Muscat AM, Nair S, Peters KB, Randazzo D, Sampson JH, Vlahovic G, Harrison WT, McLendon RE, Ashley D, Bigner DD. Recurrent glioblastoma treated with recombinant poliovirus. N Engl J Med 2018;379(2):150–61.
[48] Thirukkumaran C, Morris DG. Oncolytic viral therapy using reovirus. Methods Mol Biol 2015;1317:187–223.
[49] Bourhill T, Mori Y, Rancourt DE, Shmulevitz M, Johnston RN. Going (Reo)viral: factors promoting successful reoviral oncolytic infection. Viruses 2018;10(8).

[50] Kicielinski KP, Chiocca EA, Yu JS, Gill GM, Coffey M, Markert JM. Phase 1 clinical trial of intratumoral reovirus infusion for the treatment of recurrent malignant gliomas in adults. Mol Ther 2014;22(5):1056–62.

[51] Perez OD, Logg CR, Hiraoka K, Diago O, Burnett R, Inagaki A, Jolson D, Amundson K, Buckley T, Lohse D, Lin A, Burrascano C, Ibanez C, Kasahara N, Gruber HE, Jolly DJ. Design and selection of Toca 511 for clinical use: modified retroviral replicating vector with improved stability and gene expression. Mol Ther 2012;20(9):1689–98.

[52] Cloughesy TF, Landolfi J, Vogelbaum MA, Ostertag D, Elder JB, Bloomfield S, Carter B, Chen CC, Kalkanis SN, Kesari S, Lai A, Lee IY, Liau LM, Mikkelsen T, Nghiemphu P, Piccioni D, Accomando W, Diago OR, Hogan DJ, Gammon D, Kasahara N, Kheoh T, Jolly DJ, Gruber HE, Das A, Walbert T. Durable complete responses in some recurrent high-grade glioma patients treated with Toca 511 + Toca FC. Neuro-Oncology 2018;20(10):1383–92.

[53] Anon. Tocagen reports results of toca 5 phase 3 trial in recurrent Brain cancer, http://ir.tocagen.com/news-releases/news-release-details/tocagen-reports-results-toca-5-phase-3-trial-recurrent-brain; September 12, 2019. [Accessed 6 October 2019].

[54] Haddad D. Genetically engineered vaccinia viruses as agents for cancer treatment, imaging, and transgene delivery. Front Oncol 2017;7:96.

[55] Foloppe J, Kempf J, Futin N, Kintz J, Cordier P, Pichon C, Findeli A, Vorburger F, Quemeneur E, Erbs P. The enhanced tumor specificity of TG6002, an armed oncolytic vaccinia virus deleted in two genes involved in nucleotide metabolism. Mol Ther Oncolytics 2019;14:1–14.

CHAPTER

5

Adoptive cell therapy for glioma

Lan B. Hoang-Minh and Duane A. Mitchell

Lillian S. Wells Department of Neurosurgery, UF Brain Tumor Immunotherapy Program, Preston A. Wells, Jr. Center for Brain Tumor Therapy, McKnight Brain Institute, University of Florida, Gainesville, FL, United States

Abstract

Despite aggressive surgery, radiation, and systemic chemotherapy, the prognosis for patients diagnosed with malignant brain tumors remains dismal, and there is an urgent need for the development of treatments that are more effective and will add minimal toxicity to standard therapies. Cancer immunotherapy is a modality that holds promise for the delivery of tumor-specific cytotoxicity, with the potential to eliminate brain tumor cells without harming the eloquent brain. Among cancer immunotherapies, immune checkpoint inhibition has revolutionized cancer treatment, but its efficacy is limited to tumors that contain sufficient numbers of endogenous tumor-specific T cells that can be re-stimulated to recognize their targets and mediate tumor killing. For tumors in which tumor-reactive T cells are absent or fail to function, adoptive cellular therapy has been developed to enhance the population of antitumor effector cells. Adoptive immunotherapy involves the ex vivo expansion of autologous tumor-specific immune cells isolated from cancer patients' resected tumor tissue or peripheral blood followed by the infusion of large numbers of these expanded effector cells back into the patients. These immune cells can be administered either locally at the tumor sites or systemically via intravenous route. Formerly, immune cells with undetermined antigen specificity, such as lymphokine-activated killer cells, were predominantly used. Recently, antigen-targeted approaches have been developed, primarily involving the use of chimeric antigen receptor and T-cell receptor-transduced cells. However, although these approaches have been quite successful for other cancers, they present with unique challenges when applied to malignant brain tumors. We review the adoptive immunotherapeutic strategies that have been investigated in the treatment of glioma, including nonspecific cell therapies and tumor-specific therapies using tumor-infiltrating lymphocytes, genetically engineered T cells, and in vitro-stimulated antigen-specific T cells.

Abbreviations

ACT	adoptive cellular therapy
ATCT	adoptive T-cell transfer
BBB	blood-brain barrier
CAR	chimeric antigen receptor
CMV	cytomegalovirus
CNS	central nervous system

Immunotherapeutic Strategies for the Treatment of Glioma
https://doi.org/10.1016/B978-0-12-819755-4.00011-4

CRS	cytokine release syndrome
CTLs	cytotoxic lymphocytes
DC	dendritic cell
DI-TMZ	dose-intense temozolomide
EGFRvIII	epidermal growth factor variant 3
EphA2	ephrin type-A receptor 2
GBM	glioblastoma
GMCSF	granulocyte macrophage colony-stimulating factor
gp100	glycoprotein 100
Gy	Gray (unit)
HER2	human epidermal growth factor receptor 2
HSPCs	hematopoietic stem and progenitor cells
IDH1	isocitrate dehydrogenase 1
IL-13Rα2	interleukin 13 receptor alpha 2
IL-15	interleukin 15
IL-6	interleukin 6
IL-7	interleukin 7
LAK cells	lymphokine-activated killer cells
MAGE-A1	melanoma-associated antigen A1
MAGE-A3	melanoma-associated antigen A3
mOS	median overall survival
NK	natural killer cells
PBMCs	peripheral blood mononuclear cells
rIL-2	recombinant interleukin 2
RNA	ribonucleic acid
SOX2	sex-determining region Y-box 2
TBI	total body irradiation
TCR	T-cell receptor
TILs	tumor-infiltrating lymphocytes
TRP2	tyrosinase-related protein 2
WT1	Wilms Tumor 1

Conflict of interest

No potential conflicts of interest were disclosed.

Introduction

With the success of checkpoint inhibitor immunotherapy against hematologic malignancies and solid tumors (melanoma and carcinomas of the lung, breast, prostate, bladder, and kidney), various therapeutic strategies targeting the immune system have been attempted for brain tumors. The rationale for using immunotherapies against these tumors has been further strengthened by a revision in our understanding of the "immune-privileged" nature of the CNS, based on the landmark discovery of a functional lymphatic system surrounding the dura mater of the brain [1], improved understanding of effector T-cell trafficking in the CNS [2], as well as studies showing that the BBB within brain tumors is often disrupted [3], thus leading to enhanced access for immune effector cells and molecules. Furthermore, while the CNS parenchyma is largely devoid of circulating resting lymphocytes, activated T cells readily cross the BBB and survey brain tissue as part of their surveillance function against CNS infections [4]. Brooks and Roszman were pioneers in investigating the interactions between GBM and the immune system in the early 1970s [5, 6]. Ensuing studies tested various immunotherapeutic strategies targeting GBM in the 1980s [7–10].

Active immunotherapeutic approaches, such as the vaccination of patients or administration of checkpoint inhibitors, rely on the in vivo activation and production of adequate numbers of endogenous and functional antitumor immune cells. In contrast, the adoptive transfer of ex vivo-activated cytotoxic effector cells or ACT, a form of passive immunotherapy, allows for the ex vivo generation of large numbers of autologous tumor-specific immune cells that can be reintroduced into the patient. These effector cells have mainly been composed of T cells but can also involve NK cells, NK T cells, or γδ T cells. For glioma patients, these cells are isolated from patients' peripheral blood, tumor-draining lymph nodes, or resected tumor tissue; these cells are then activated, expanded ex vivo, and reintroduced in large numbers (10^6–10^9) into the patients, either systemically or intracranially. While T cells are being expanded, they can be selected for tumor specificity and/or increased effector function. They can also be modified to express TCRs for tumor-associated/specific antigen peptides or CARs for tumor cell surface proteins. GBM-associated antigens that have been examined so far include IL-13Rα2, HER2, gp100, EphA2, survivin, WT1, TRP2, SOX2, MAGE-A1, MAGE-A3, and CMV. GBM-specific targets include EGFRvIII variant and the IDH-1 mutant (R132H) [11]. Table 1 summarizes the recent published findings from clinical trials of those various ACT approaches for glioma.

Strategies for adoptive cell therapy for glioma

Nonspecific cell therapies

Lymphokine-activated killer cells

LAK cells are derived from autologous PBMCs activated with rIL-2 in vitro [9] and comprised of a major fraction of nonspecific activated T cells and a minor fraction of activated NK cells. In 1985, a study by Rosenberg et al. involving the administration of autologous LAK cells along with rIL-2 in 25 patients with metastatic malignant growth refractory to standard therapies showed ~50% regression of the tumor in 11 out of 25 treated patients. Complete tumor regression that lasted for 10 months occurred in one patient with metastatic melanoma. Partial response was observed in nine patients who suffered from hepatic or pulmonary metastases from melanoma and other cancers [24]. In another study, comparison of LAK cell therapy and IL-2 with IL-2 treatment alone showed no difference in response rates in patients with renal cell carcinoma [25]. Moreover, high doses of IL-2 led to serious adverse reactions such as capillary leak syndrome (hypotension, oliguria, pulmonary edema, and dyspnea), halting further clinical investigations.

Several studies have demonstrated the safety and efficacy of focal LAK cell therapy in patients with malignant gliomas since the 1980s [8–10, 26–31], with some that showed promising results in prolonging disease-free survival. In 2001, Hayes et al reported that 6 out of 28 patients with recurrent high-grade primary brain tumors achieved long-term survival (over 2 years) following resection and local injection of LAK cells and low doses of IL-2 [28]. Another study verified the safety and feasibility of focal LAK therapy in 40 GBM patients and also showed increased mOS of 17.5 months for 31 patients treated with

TABLE 1 Published trials with ACT for glioma.[a]

Title/trial NCT	Sponsor	Phase	*n*	OS	PFS	Status	Ref.
Evaluation of Recovery from Drug-Induced Lymphopenia Using Cytomegalovirus-specific T-cell Adoptive Transfer/ NCT00693095	John Sampson	1	23	NR	NR	Completed	[12]
Phase I Study of Cellular Immunotherapy Using Memory Enriched T Cells Lentivirally Transduced to Express an IL-13Rα2-Specific, Hinge-Optimized, 41BB-Costimulatory Chimeric Receptor and a Truncated CD19 for Patients with Recurrent/Refractory Malignant Glioma/NCT02208362	City of Hope Medical Center	1	135	NR	NR	Recruiting	[13]
Multi-center, Randomized, Open-label Phase 3 Clinical Trial to Assess the Efficacy and Safety of "INNOCELL Immuncell-LC" With Temozolomide in Newly Diagnosed Glioblastoma of Korea/NCT00807027	Green Cross Cell Corporation	3	180	22.5 months	8.1 months	Completed	[14–16]
Pilot Feasibility and Safety Study of Cellular Immunotherapy for Recurrent/Refractory Malignant Glioma Using Genetically Modified Autologous CD8-positive T Cell Clones/NCT00730613	City of Hope Medical Center	1	3	11 months post relapse	NR	Completed	[17–19]
Phase I Study of Cellular Immunotherapy for Recurrent/ Refractory Malignant Glioma Using Intratumoral Infusions of GRm13Z40-2, An Allogeneic CD8-positive Cytolytic T-Cell Line Genetically Modified to Express the IL 13-Zetakine and HyTK and to be Resistant to Glucocorticoids, in Combination with Interleukin-2/NCT01082926	City of Hope Medical Center	1	6	NR	NR	Completed	[20]
Phase I Trial to Assess Safety of Autologous Human Cytomegalovirus (HCMV)-specific T Cell Therapy for Glioblastoma Multiforme/ACTRN12609000338268	Queensland Institute of Medical Research	1	19	13.3 months	8.2 months	Completed	[21]
Administration of HER2 Chimeric Antigen Receptor Expressing CMV-Specific Cytotoxic T Cells in Patients with Glioblastoma Multiforme/NCT01109095	Baylor College of Medicine	1	16	24.5 months	3.5 months after T cell infusion	Active, not recruiting	[22]
A Controlled, Randomized, Parallel Group, Multicenter Study of the Efficacy and Safety of Herpes Simplex Virus-Thymidine Kinase Gene Therapy (CereproTM), with Subsequent Ganciclovir, for the Treatment of Patients with Operable High-Grade Malignant Glioma/EudraCT # 2004-000464-28	Ark Therapeutics Limited	3	250	10.3 months	NR	Completed	[23]

[a] *Worldwide trials reported within the past 7 years.*

Note: Most study titles and identifiers obtained from clinicaltrials.gov.

n, number of participants; *OS*, overall survival (median); *PFS*, progression-free survival (median); NR, not reported.

this immunotherapy, versus 13.6 months for a control group of 41 GBM patients [26]. Several groups subsequently tested modified LAK cell therapy protocols. Although some of these clinical studies showed partial tumor response in some recurrent GBM patients following focal injection of LAK cells and high-dose IL-2, they also reported neurological side effects and several toxicities, including cerebral edema, with no significant differences in mOS [32–34].

NK cells constitute the major effector population among LAK cells and are able to recognize cancer cells in a non-MHC-restricted manner. As such, they may represent a primitive immune surveillance system that is capable of recognizing and destroying abnormal cells and might play an important role in cellular immune responses in cooperation with DCs, CTLs, and other cell types. The extent of NK cells' cytotoxic activity against malignant tumor cells is unknown [35]. The function of NK cells might be complementary to CTLs' activity against tumor cells that is dependent on antigen recognition through MHC class I molecules. NK cells for adoptive immune therapy can be derived from hematopoietic stem cells, induced pluripotent stem cells, genetically modified NK cells, allogeneic and autologous NK cells, as well as NK cell lines [36].

The clinical application of NK cell therapy has been reported in only a limited number of cases in malignant brain tumors or other neoplasms [37, 38]. A major hurdle has been the enrichment and large-scale expansion of these cells. However, following advances in the field of NK cell biology [39], a renewed interest in NK cell-based immunotherapy for cancer emerged [40]. Preclinical advancements of NK cell adoptive immunotherapy have included the manipulation of receptor-mediated activation, amplification of antibody-dependent cellular cytotoxicity, and transfer of ex vivo-expanded, CAR-engineered NK cells (reviewed in [41]). Recently, Harada et al. reported a simple method for NK cell expansion [42], which was followed by a clinical trial involving NK cell therapy for patients with recurrent malignant gliomas [38]. NK cells were expanded from autologous PBMCs by co-culture with an irradiated human feeder cell line (Wilms' tumor cell line) using rIL-2, before being injected intravenously and/or intracranially into nine patients. The clinical evaluation demonstrated a partial or minor response in half of the patients and no change or progressive disease in the other half after a total of 16 courses of treatment. None of the patients experienced severe neurological toxicity.

LAK therapy for glioma faces several limitations. First, the mechanisms by which these cells exert their cytotoxic activity on tumor cells are still unknown. Additionally, LAK cells do not specifically home to brain tumors, and although locoregional delivery does not represent a major limitation, as most brain tumors have not been observed to metastasize outside of the CNS, the focus of adoptive immunotherapy for brain tumors has shifted to tumor-specific T-cell lymphocytes with multiple log-fold higher cytotoxic activity.

Tumor-specific cell therapies

ATCT has become a promising option for cancer treatment, as the basic conditions for its technical feasibility have been achieved in recent years, and this modality presents with several advantages compared with other forms of immunotherapy. The discovery that IL-2 can promote T-cell expansion ex vivo without affecting the functionality of these effector cells [43], along with other advances in cell culture, has made ATCT technically feasible, as sufficient quantities of human T cells can be generated for subsequent infusion. T cells with chosen specificities and improved functionality for potent antitumor responses can be selected and grown in vitro, therefore avoiding adverse reactions in vivo. Importantly, the tumor microenvironment can be manipulated to make the tumors more receptive to ATCT, e.g., by blocking mechanisms of immunosuppression through reducing T-regulatory lymphocytes [44]. Cell sources used in ATCT for glioma include TILs, genetically engineered T cells, and in vitro-stimulated tumor antigen-specific T cells.

Tumor-infiltrating lymphocytes

TILs consist of immune cells that have been extracted from fresh cancerous tissue. TILs have attracted much attention as candidates for cancer immunotherapy following tumor resection as they have presumably undergone in vivo selection for their tumor recognition abilities. For this adoptive immune therapy, TILs are isolated from the tumor, selected, activated, and expanded ex vivo using IL-2. A preparative lymphodepleting regimen accompanied by TBI is sometimes administered to suppress endogenous regulatory T cells and other immune-suppressive cells, thus rendering the environment suitable for efficient TILs' antitumor activity. Ex vivo-expanded TILs are then re-infused into the patient [45].

The application of TILs in cancer treatment was first established by Rosenberg and colleagues in patients with refractory metastatic melanoma [46]. Patients were treated with autologous TILs along with IL-2 after a lymphodepleting preparative regimen (using cyclophosphamide or fludarabine). Some patients also received either 2- or 12-Gy TBI. The response rate to this treatment course ranged from 49% to 72% and was positively correlated with the degree of lymphodepletion [47]. Some groups have reported a few cases treated with a combination of lymphodepleting, high-dose chemotherapy, followed by adoptive TIL transfer for metastatic brain tumors. Dudley et al. reported a combination of lymphodepleting chemotherapy followed by the adoptive transfer of TILs for the treatment of 35 patients with refractory metastatic melanoma [48]. In this study, 18 patients, including a patient with brain metastases, experienced objective clinical responses. In addition, removing suppressor T cells or reducing endogenous lymphocytes that can compete for homeostatic regulatory cytokines, such as IL-15 or IL-7, results in lymphodepletion of the host and enhancement of TILs' function [49, 50]. Despite these advantages, ACT with TILs presents some limitations. IL-2 stimulates the expansion of regulatory T cells, which can, in turn, down-regulate the antitumor response. Although the use of TILs is of highest efficiency against melanoma when compared with different cancer types, 50% of melanomas do not produce adequate antitumor TILs [45, 51], which might be related to tumor size, location, and/or immunogenicity [52]. In addition, the process of ACT with TILs is lengthy and expensive [45].

While the adoptive transfer of autologous TILs has produced some durable objective responses in patients with advanced melanoma, the application of this therapy has proved to be challenging in early clinical trials for malignant glioma due to the low density of TILs within tumors, the deterioration of their function ("exhaustion"), and their limited tumor-selective cytotoxicity [53–55]. One pilot study by Quattrocchi et al., which involved the local injection of autologous TILs combined with low-dose adjuvant chemotherapy in six patients with recurrent malignant glioma, reported one complete response, two partial responses, and three cases of progressive disease [56]. In subsequent studies, patients were first vaccinated with irradiated autologous tumor cells before receiving adoptive transfer of stimulated T cells isolated from their peripheral blood or draining lymph nodes. Although safe, these studies did not show significant OS benefits compared with control groups.

An important factor in selecting TILs that have optimal antitumor activity could be their telomere length, which is correlated with the persistence of T cells and clinical response after ACT in melanoma patients [57, 58], suggesting that less-proliferated TILs are more effective for ACT. Thus, telomere length could be used as a marker to select TILs to improve the efficacy of ACT. The persistence of antitumor T-cell clones is responsible for tumor regression in patients with metastatic melanoma receiving ACT with TILs [14]. In addition, it has been shown that IL-15 and IL-7 are more effective than IL-2 in preventing regulatory T-cell proliferation [59]. Chronic depletion of regulatory T cells using anti-CD25 monoclonal daclizumab has also been demonstrated [60–62].

CAR T cells

To overcome limitations such as finding a suitable subpopulation of tumor-specific lymphocytes among the heterogeneous population of TILs or PBMCs, genetically designed T lymphocytes with specific antitumor activity have been developed for use in ACT for patients with malignant growth [62]. Retroviral and lentiviral vectors, and more recently the Sleeping Beauty transposon system, carrying CAR or TCR genes that are specific for tumor antigens are transduced into peripheral blood T cells leading to the production of CAR T cells and TCR T cells, respectively [63–67]. Although TCR-modified T cells have been extensively studied in patients with advanced melanoma [68] and conferred an OS advantage in a murine xenograft model of pediatric glioma expressing the H3.3 K27M mutation [20], no clinical studies have investigated the efficacy of TCR-modified T cells against malignant glioma so far. In order to develop effective TCR-based therapeutic approaches targeting malignant glioma, novel tumor-specific neoantigens will need to be identified. Furthermore, methods to overcome inefficient expression of transduced TCRs in T lymphocytes and self-reactivity due to mispairing of introduced TCR chains with endogenous TCR subunits will need to be developed [69, 70].

First-generation CAR T cells are genetically engineered to display a CAR that includes an extracellular single-chain variable fragment of an antibody permitting the specific recognition of a tumor cell's surface protein. This antibody fragment is bound to the intracellular signaling domains of CD3ζ, which allows for T-cell activation and cytotoxic function upon tumor protein binding. Second- and third-generation CARs possess signal transduction domains of CD28 and additional intracellular costimulatory domains, such as OX40, 4-1BB. These genetically engineered lymphocytes have high tumor reactivity and avidity [54, 71].

Compared with TCR-modified T cells, CAR T cells are able to recognize a broader range of tumor antigen targets that do not have to undergo processing for effective display on MHC molecules; they can also avoid immune evasion mechanisms, whereby tumors down-regulate MHC expression, and can be used in patients irrespective of their HLA type. Particularly, CAR T cells have demonstrated considerable success in the treatment of hematologic malignancies that are refractory to standard chemotherapy [14, 67]. Thus, considerable efforts have been undertaken in applying this therapeutic approach to solid tumors. A phase I study in recurrent GBM patients, which examined surgically resected tumor samples following the intravenous administration of CAR T cells targeting EGFRvIII, has shown that those

CAR T cells are able to traffic to the tumors, proliferate in situ, and exert anti-EGFRvIII activity that leads to the loss of EGFRvIII expression in those tumors [72, 73]. T-cell repertoire screening identified a significant increase in the number and clonotypic diversity of TILs following CAR T-cell infusion, possibly due to EGFRvIII CAR T-cell antitumor activity and subsequent epitope spreading. CD8-positive effector T cells, other activated immune cells, and the expression of granzyme B, IFN-γ, and CD25, were also increased in tumors post CAR T-cell infusion compared with preinfusion. However, compensatory immunosuppressive molecules such as IDO-1, PD-L1, TGF-β, IL-10, and Foxp3 also increased following CAR T-cell infusion. Nonetheless, EGFRvIII-CAR T treatment induced an immunogenic tumor microenvironment without apparent neurotoxicity [72, 74].

IL-13Rα2 is a monomeric high-affinity IL-13 receptor that is overexpressed by more than half of GBM tumors and is a prognostic indicator of poor patient survival [75]. A phase I trial (NCT02208362) has demonstrated the tolerability of IL-13Rα2-specific CAR T-cell administration into the resection cavity in three patients with recurrent GBM, with evidence of transient antiglioma responses such as reduced IL-13Rα2 expression within the tumor and increase in tumor necrotic volume at the site of CAR T-cell administration [76–79]. Notably, a recent case report of a patient with recurrent multifocal GBM describes dramatic widespread tumor regressions that followed local and intraventricular infusions of autologous anti-IL-13Rα2 CAR T cells, along with corresponding increases in cytokine levels and immune cells in their cerebrospinal fluid, which continued for several months after the initiation of CAR T-cell therapy [23]. The limitations of TCR-modified and CAR-based strategies include the identification of tumor-specific antigens and the toxicities due to off-tumor targeting of normal tissue. Antigen escape also represents a common resistance mechanism, whereby tumor cells lacking the target antigen mediate disease progression and recurrence. Therefore, the heterogeneity of tumor antigen expression and tumor immunosuppressive mechanisms remain major obstacles to the efficacy of CAR T-cell therapy for glioma but may potentially be overcome by combinatorial approaches targeting multiple antigens and the immune-suppressive environment.

In vitro-stimulated antigen-specific T cells

Among cells associated with acquired immunity, CTLs have been shown to play a crucial role in tumor rejection in vivo. CTLs are CD8-positive T cells that are able to eradicate tumor cells in a MHC class I-restricted manner. Numerous preclinical studies have shown that immune therapy with CTLs is effective against gliomas. In addition, tumor-specific CD4-positive T lymphocytes demonstrate cytotoxic activity against tumor cells but also play an important role in CTL-mediated tumor cell killing [80]. Therefore, adoptive immunotherapy strategies using autologous tumor-specific T lymphocytes consisting of CD8-positive and CD4-positive cells have been tested in patients with malignant gliomas. A number of clinical trials using CTLs or autologous tumor-specific T lymphocytes have been reported. A study by Kruse et al. reported a decrease in tumor growth in three out of five patients with malignant gliomas treated with intracavitary alloreactive CTLs and IL-2 [43]. In 1996, Tsurushima et al. demonstrated a method to expand autologous tumor-specific T lymphocytes ex vivo from PBMCs using autologous GBM cells, without preceding injection of tumor vaccine [81]. A clinical pilot study involving intracranial or intrathecal injections of autologous tumor-specific T lymphocytes expanded using this method showed 1 complete response and 4 partial responses in 10 cases with high-grade gliomas, without intolerable toxicities [44, 82]. Combinations of autologous irradiated tumor vaccine and specific cell therapy using LAKs/ autologous tumor-specific T lymphocytes isolated from PBMCs or lymph nodes were also attempted on patients with malignant gliomas [83–86]. Plautz et al. reported partial responses in four out of eight cases with residual tumor, among 12 cases of newly diagnosed gliomas treated with LAK/autologous tumor-specific T lymphocytes (71% CD4-positive T cells) expanded from inguinal lymph nodes after the injection of autologous irradiated tumor vaccine [84].

The ubiquitous expression of human CMV antigens on GBM cells has provided a unique opportunity to target those viral antigens in adoptive immunotherapy for GBM. Patient-derived T cells stimulated by CMV pp65 RNA-transfected DCs can effectively target and eliminate autologous GBM tumor cells in an antigen-specific manner, further supporting the rationale for the development of CMV-directed immunotherapy for GBM patients [87–89]. A randomized phase I pilot trial in 22 CMV-seropositive, newly diagnosed GBM patients investigated the use of ACT with CMV pp65-specific T cells generated ex vivo using autologous CMV pp65 RNA-transfected DCs, with or without CMV pp65-DC vaccination (NCT00693095). Prospective results of this study suggested that increases in polyfunctional CMV-specific CD8-positive T cells (which are able to simultaneously secrete IFN-γ, TNF-α, and CCL3) correlated with mOS, and thus, that ex vivo analyses of T-cell polyfunctionality might help predict patient responses following immunotherapy. In addition to further reinforcing that CMV pp65 is a robust immunogenic GBM antigen, this study was also the first to report that DC vaccination enhances the polyfunctionality of transferred CMV-specific CD8-positive T cells in vivo in patients with newly diagnosed GBM [22]. Another early-phase trial conducted in Australia in 10 patients with recurrent GBM who received autologous CMV-specific T-cell therapy resulted in a median OS of 403 days. Notably, the T cells used for adoptive immunotherapy displayed distinct gene expression signatures, particularly for checkpoint inhibitory receptors and T-cell transcription factors, which correlated with the patient's clinical responses [90]. Although further assessments with larger cohorts of patients are warranted, these studies suggest that adoptive immunotherapy using CMV-specific T cells is safe and may provide long-term clinical benefits for patients with newly diagnosed or recurrent GBM.

Furthermore, our group has shown that intravenously administered HSPCs traffic to sites of malignant glioma growth and significantly enhance the efficacy of ACT in preclinical studies [91–93]. ACT incorporating HSPC transfer is effective against multiple preclinical brain tumor models (GBM, diffuse intrinsic brain stem glioma, and Group 3 medulloblastoma). Transferred HSPCs supplant host myeloid-derived suppressor cells, differentiate into potent APCs, and present captured tumor antigens, thus leading to the enhancement of intratumoral T-cell activation and promoting the immunologic rejection of invasive gliomas [92]. A phase I/II clinical trial in patients with recurrent medulloblastoma or primitive neuroectodermal tumors evaluated the safety and efficacy of ACT concurrent with total tumor RNA-loaded DC vaccination and autologous HSPC transplantation during recovery from myeloablative chemotherapy (NCT01326104). In one patient with relapsed medulloblastoma receiving ACT, an early and massive expansion of tumor-reactive lymphocytes, coupled with prolonged persistence in the peripheral blood, was observed, along with long-term progression-free survival (13 months), compared with a historical patient cohort of recurrent medulloblastoma patients treated with myeloablative chemotherapy and stem cell rescue alone [94]. In addition, based on the results of studies showing the effectiveness of adjuvant Td preconditioning, two phase I clinical trials have investigated the feasibility and safety of ACT concurrent with total tumor RNA-DC vaccine/GM-CSF and Td booster vaccine for the treatment of newly diagnosed brain stem gliomas during recovery from focal radiotherapy, with or without DI-TMZ (NCT03396575), and the treatment of pediatric high-grade gliomas with autologous HSPC transfer and DI-TMZ (NCT03334305).

Challenges and future directions

Effective and safe ACT therapy may represent a promising modality that turns "cold" GBM tumors into "hot" environments. Two critical questions related to ACT are whether new T-cell sources can be developed to replace autologous cell production and overcome histocompatibility barriers, and which would be the best method to minimize potential on-target/off-tumor toxicities. Initial clinical studies of CAR-T therapy have shown promising results for the treatment of advanced glioma. Induction of robust inflammatory responses in glioma by intravenous infusion of EGFRvIII CAR-T therapy suggests that it is possible to turn these tumors into "hot" environments, without inducing neurotoxicity. However, although none of the CAR-T trials for GBM have demonstrated significant neurotoxicity to date, CD19 CAR-T therapy studies for pediatric acute lymphoblastic leukemia have resulted in severe neurotoxicity and treatment-related fatalities [95]. This toxicity was associated with cytokine-release syndrome (CRS), BBB permeability, and endothelial activation [74, 96–98]. In a pilot study, pretreatment with a combination of anti-IL-6 monoclonal antibody and dexamethasone abrogated the rates of severe CRS, without affecting mOS or overall response rate [99]. Other strategies that have been used to avoid CRS-related neurotoxicity include locoregional delivery of CAR T cells, limiting systemic immune effects [23, 79, 100]. Thus, alternative routes of delivery, e.g., intraventricular, of T cells for CAR-T or other adoptive cell therapies for glioma could reduce toxicities/enhance treatment efficacy and need to be explored.

Other issues in ACT include a lack of antigen specificity due to the high, variable mutational load and heterogeneous nature of GBM, poor T-cell trafficking, as well as hostile and immune suppressive microenvironment. Different strategies are being investigated to overcome these hurdles, such as targeting multiple antigens to increase specificity and response, combining ACT with immune checkpoint inhibitors or other immunotherapies, intraventricular infusion for bypassing the systemic circulation and direct effect, and use of third-generation and fourth-generation CAR T cells in CAR-T therapy. Although ACT strategies have been successful against hematologic cancers and melanoma, they have faced challenges against other solid tumors, particularly in locations such as the brain [101]. One contributing factor lies in the poor trafficking and persistence of T cells at the tumor sites following systemic delivery [12]. ACT has emerged as the most effective treatment strategy against advanced malignant melanoma, eliciting remarkable objective clinical responses in up to 75% of patients with refractory metastatic disease [13]. Immunologic surrogate endpoints that correlate with treatment outcome have been identified in these patients, with clinical responses being dependent on the localization of transferred T cells to sites of invasive tumor growth [14, 15, 57]. These findings underscore the need for novel approaches to enhance and noninvasively monitor the trafficking and persistence of adoptively transferred immune cells to brain tumor sites. Recently, several groups have reported strategies to longitudinally track immune cells in vivo following immunotherapy, e.g., using positron emission tomography imaging or magnetic resonance imaging [102–106].

T-cell therapy targeting shared glioma neoantigens, which are derived from tumor-specific protein-coding mutations exempt from central tolerance, could help achieve effective antitumor immunity. An early-phase clinical study that used personalized neoantigen-targeting vaccines to immunize patients with newly diagnosed GBM demonstrated the generation of circulating polyfunctional, neoantigen-specific CD4-positive and CD8-positive T-cell responses that were enriched in a memory phenotype and an increase in the number of tumor-infiltrating T cells [107]. This study showed that neoantigen-specific T cells could migrate from the peripheral blood into intracranial GBM tumors. Chheda et al. recently identified a novel neoantigen epitope encompassing the K27M mutation within the histone 3 variant H3.3 that is present in a majority of pediatric diffuse midline gliomas [20, 101]. Furthermore, they cloned the cDNA for the TCR α- and β-chains derived from a high-avidity H3.3.K27M-specific CTL clone, allowing the development of novel vaccine- and TCR-transduced T-cell-based immunotherapy strategies in patients with H3.3K27M-positive gliomas. These studies suggest the importance and feasibility of using neoantigens as targets of antitumor-specific T-cell responses in the context of ACT.

Conclusions

Although the ability of adoptively transferred T cells to migrate and mediate tumor regression in areas of invasive GBM is still unclear, early clinical studies suggest that ACT may be an effective strategy in the treatment of malignant gliomas, particularly in combination with other immunotherapeutic modalities, such as DC vaccination. However, despite enthusiasm for adoptive immunotherapy in cancer, several obstacles must be addressed before ACT becomes a standard treatment for the management of brain tumors. For gliomas, which comprise multiple tumor-associated antigens, the question of which are the optimal targets to minimize tumor escape via antigen loss/down-regulation, while combating tumor heterogeneity and improving patient safety, also needs to be addressed. The requirement for nonmyeloablative preconditioning must be evaluated, as there might be a benefit to altering endogenous lymphoid populations in anticipation of their activation or recruitment by adoptively transferred T cells. Efforts also need to be directed at understanding the interactions between tumor microenvironment and endogenous or adoptively transferred immune cells. Moreover, optimal scheduling and route of cell administration still need to be determined. Knowledge of how to best minimize these toxicities, coupled with rational combinations with standard therapies and immune modulators, will pave the way for successful application of ACT strategies for the treatment of glioma.

References

[1] Louveau A, Smirnov I, Keyes TJ, Eccles JD, Rouhani SJ, Peske JD, et al. Structural and functional features of central nervous system lymphatic vessels. Nature 2015;523(7560):337–41. https://doi.org/10.1038/nature14432.

[2] Korn T, Kallies A. T cell responses in the central nervous system. Nat Rev Immunol 2017;17(3):179–94. https://doi.org/10.1038/nri.2016.144.

[3] Arvanitis CD, Ferraro GB, Jain RK. The blood-brain barrier and blood-tumour barrier in brain tumours and metastases. Nat Rev Cancer 2020;20(1):26–41. https://doi.org/10.1038/s41568-019-0205-x.

[4] Manglani M, McGavern DB. New advances in CNS immunity against viral infection. Curr Opin Virol 2018;28:116–26. https://doi.org/10.1016/j.coviro.2017.12.003.

[5] Brooks WH, Netsky MG, Normansell DE, Horwitz DA. Depressed cell-mediated immunity in patients with primary intracranial tumors. Characterization of a humoral immunosuppressive factor. J Exp Med 1972;136(6):1631–47. https://doi.org/10.1084/jem.136.6.1631.

[6] Brooks WH, Roszman TL, Rogers AS. Impairment of rosette-forming T lymphocytes in patients with primary intracranial tumors. Cancer 1976;37(4):1869–73. https://doi.org/10.1002/1097-0142(197604)37:4<1869::aid-cncr2820370435>3.0.co;2-q.

[7] Mahaley MS, Bigner DD, Dudka LF, Wilds PR, Williams DH, Bouldin TW, et al. Immunobiology of primary intracranial tumors. Part 7: active immunization of patients with anaplastic human glioma cells: a pilot study. J Neurosurg 1983;59(2):201–7. https://doi.org/10.3171/jns.1983.59.2.0201.

[8] Jacobs SK, Wilson DJ, Kornblith PL, Grimm EA. Interleukin-2 and autologous lymphokine-activated killer cells in the treatment of malignant glioma. Preliminary report. J Neurosurg 1986;64(5):743–9. https://doi.org/10.3171/jns.1986.64.5.0743.

[9] Merchant RE, Grant AJ, Merchant LH, Young HF. Adoptive immunotherapy for recurrent glioblastoma multiforme using lymphokine activated killer cells and recombinant interleukin-2. Cancer 1988;62(4):665–71. https://doi.org/10.1002/1097-0142(19880815)62:4<665::aid-cncr2820620403>3.0.co;2-o.

[10] Yoshida S, Tanaka R, Takai N, Ono K. Local administration of autologous lymphokine-activated killer cells and recombinant interleukin 2 to patients with malignant brain tumors. Cancer Res 1988;48(17):5011–6.

[11] Fecci PE, Heimberger AB, Sampson JH. Immunotherapy for primary brain tumors: no longer a matter of privilege. Clin Cancer Res 2014;20(22):5620–9. https://doi.org/10.1158/1078-0432.CCR-14-0832.

[12] Slaney CY, Kershaw MH, Darcy PK. Trafficking of T cells into tumors. Cancer Res 2014;74(24):7168–74. https://doi.org/10.1158/0008-5472.CAN-14-2458.

[13] Phan GQ, Rosenberg SA. Adoptive cell transfer for patients with metastatic melanoma: the potential and promise of cancer immunotherapy. Cancer Control 2013;20(4):289–97. https://doi.org/10.1177/107327481302000406.

[14] Robbins PF, Dudley ME, Wunderlich J, El-Gamil M, Li YF, Zhou J, et al. Cutting edge: persistence of transferred lymphocyte clonotypes correlates with cancer regression in patients receiving cell transfer therapy. J Immunol 2004;173(12):7125–30.

[15] Rosenberg SA, Restifo NP, Yang JC, Morgan RA, Dudley ME. Adoptive cell transfer: a clinical path to effective cancer immunotherapy. Nat Rev Cancer 2008;8(4):299–308. https://doi.org/10.1038/nrc2355.

[16] Powell DJ, Dudley ME, Robbins PF, Rosenberg SA. Transition of late-stage effector T cells to CD27+ CD28+ tumor-reactive effector memory T cells in humans after adoptive cell transfer therapy. Blood 2005;105(1):241–50. https://doi.org/10.1182/blood-2004-06-2482.

[17] Chapelin F, Capitini CM, Ahrens ET. Fluorine-19 MRI for detection and quantification of immune cell therapy for cancer. J Immunother Cancer 2018;6(1):105. https://doi.org/10.1186/s40425-018-0416-9.

[18] Tremblay ML, Davis C, Bowen CV, Stanley O, Parsons C, Weir G, et al. Using MRI cell tracking to monitor immune cell recruitment in response to a peptide-based cancer vaccine. Magn Reson Med 2018;80(1):304–16. https://doi.org/10.1002/mrm.27018.

[19] Wei W, Jiang D, Ehlerding EB, Luo Q, Cai W. Noninvasive PET imaging of T cells. Trends Cancer 2018;4(5):359–73. https://doi.org/10.1016/j.trecan.2018.03.009.

[20] Chheda ZS, Kohanbash G, Okada K, Jahan N, Sidney J, Pecoraro M, et al. Novel and shared neoantigen derived from histone 3 variant H3.3K27M mutation for glioma T cell therapy. J Exp Med 2018;215(1):141–57. https://doi.org/10.1084/jem.20171046.

[21] Phuphanich S, Wheeler CJ, Rudnick JD, Mazer M, Wang H, Nuño MA, et al. Phase I trial of a multi-epitope-pulsed dendritic cell vaccine for patients with newly diagnosed glioblastoma. Cancer Immunol Immunother 2013;62(1):125–35. https://doi.org/10.1007/s00262-012-1319-0.

[22] Reap EA, Suryadevara CM, Batich KA, Sanchez-Perez L, Archer GE, Schmittling RJ, et al. Dendritic cells enhance polyfunctionality of adoptively transferred T cells that target cytomegalovirus in glioblastoma. Cancer Res 2018;78(1):256–64. https://doi.org/10.1158/0008-5472.can-17-0469.

[23] Brown CE, Alizadeh D, Starr R, Weng L, Wagner JR, Naranjo A, et al. Regression of glioblastoma after chimeric antigen receptor T-cell therapy. N Engl J Med 2016;375(26):2561–9. https://doi.org/10.1056/NEJMoa1610497.

[24] Rosenberg SA, Lotze MT, Muul LM, Leitman S, Chang AE, Ettinghausen SE, et al. Observations on the systemic administration of autologous lymphokine-activated killer cells and recombinant interleukin-2 to patients with metastatic cancer. N Engl J Med 1985;313(23):1485–92. https://doi.org/10.1056/NEJM198512053132327.

[25] Law TM, Motzer RJ, Mazumdar M, Sell KW, Walther PJ, O'Connell M, et al. Phase III randomized trial of interleukin-2 with or without lymphokine-activated killer cells in the treatment of patients with advanced renal cell carcinoma. Cancer 1995;76(5):824–32. https://doi.org/10.1002/1097-0142(19950901)76:5<824::aid-cncr2820760517>3.0.co;2-n.

[26] Dillman RO, Duma CM, Schiltz PM, DePriest C, Ellis RA, Okamoto K, et al. Intracavitary placement of autologous lymphokine-activated killer (LAK) cells after resection of recurrent glioblastoma. J Immunother 2004;27 (5):398–404. https://doi.org/10.1097/00002371-200409000-00009.

[27] Hayes RL, Koslow M, Hiesiger EM, Hymes KB, Hochster HS, Moore EJ, et al. Improved long term survival after intracavitary interleukin-2 and lymphokine-activated killer cells for adults with recurrent malignant glioma. Cancer 1995;76(5):840–52. https://doi.org/10.1002/1097-0142(19950901)76:5<840::aid-cncr2820760519>3.0.co;2-r.

[28] Hayes RL, Arbit E, Odaimi M, Pannullo S, Scheff R, Kravchinskiy D, et al. Adoptive cellular immunotherapy for the treatment of malignant gliomas. Crit Rev Oncol Hematol 2001;39(1–2):31–42. https://doi.org/10.1016/s1040-8428(01)00122-6.

[29] Lillehei KO, Mitchell DH, Johnson SD, McCleary EL, Kruse CA. Long-term follow-up of patients with recurrent malignant gliomas treated with adjuvant adoptive immunotherapy. Neurosurgery 1991;28(1):16–23. https://doi.org/10.1097/00006123-199101000-00003.

[30] Okamoto Y, Shimizu K, Tamura K, Miyao Y, Yamada M, Matsui Y, et al. An adoptive immunotherapy of patients with medulloblastoma by lymphokine-activated killer cells (LAK). Acta Neurochir 1988;94(1–2):47–52. https://doi.org/10.1007/bf01406615.

[31] Sankhla SK, Nadkarni JS, Bhagwati SN. Adoptive immunotherapy using lymphokine-activated killer (LAK) cells and interleukin-2 for recurrent malignant primary brain tumors. J Neuro-Oncol 1996;27(2):133–40. https://doi.org/10.1007/bf00177476.

[32] Barba D, Saris SC, Holder C, Rosenberg SA, Oldfield EH. Intratumoral LAK cell and interleukin-2 therapy of human gliomas. J Neurosurg 1989;70(2):175–82. https://doi.org/10.3171/jns.1989.70.2.0175.

[33] Blancher A, Roubinet F, Grancher AS, Tremoulet M, Bonaté A, Delisle MB, et al. Local immunotherapy of recurrent glioblastoma multiforme by intracerebral perfusion of interleukin-2 and LAK cells. Eur Cytokine Netw 1993;4(5):331–41.

[34] Boiardi A, Silvani A, Ruffini PA, Rivoltini L, Parmiani G, Broggi G, et al. Loco-regional immunotherapy with recombinant interleukin-2 and adherent lymphokine-activated killer cells (A-LAK) in recurrent glioblastoma patients. Cancer Immunol Immunother 1994;39(3):193–7. https://doi.org/10.1007/bf01533386.

[35] Bordignon C, Carlo-Stella C, Colombo MP, De Vincentiis A, Lanata L, Lemoli RM, et al. Cell therapy: achievements and perspectives. Haematologica 1999;84(12):1110–49.

[36] Hu W, Wang G, Huang D, Sui M, Xu Y. Cancer immunotherapy based on natural killer cells: current progress and new opportunities. Front Immunol 2019;10:1205. https://doi.org/10.3389/fimmu.2019.01205.

[37] Arai S, Meagher R, Swearingen M, Myint H, Rich E, Martinson J, et al. Infusion of the allogeneic cell line NK-92 in patients with advanced renal cell cancer or melanoma: a phase I trial. Cytotherapy 2008;10(6):625–32. https://doi.org/10.1080/14653240802301872.

[38] Ishikawa E, Tsuboi K, Saijo K, Harada H, Takano S, Nose T, et al. Autologous natural killer cell therapy for human recurrent malignant glioma. Anticancer Res 2004;24(3b):1861–71.

[39] Fujisaki H, Kakuda H, Shimasaki N, Imai C, Ma J, Lockey T, et al. Expansion of highly cytotoxic human natural killer cells for cancer cell therapy. Cancer Res 2009;69(9):4010–7. https://doi.org/10.1158/0008-5472.CAN-08-3712.
[40] Barkholt L, Alici E, Conrad R, Sutlu T, Gilljam M, Stellan B, et al. Safety analysis of ex vivo-expanded NK and NK-like T cells administered to cancer patients: a phase I clinical study. Immunotherapy 2009;1(5):753–64. https://doi.org/10.2217/imt.09.47.
[41] Rezvani K, Rouce RH. The application of natural killer cell immunotherapy for the treatment of cancer. Front Immunol 2015;6:578. https://doi.org/10.3389/fimmu.2015.00578.
[42] Harada H, Saijo K, Watanabe S, Tsuboi K, Nose T, Ishiwata I, et al. Selective expansion of human natural killer cells from peripheral blood mononuclear cells by the cell line, HFWT. Jpn J Cancer Res 2002;93(3):313–9. https://doi.org/10.1111/j.1349-7006.2002.tb02174.x.
[43] Kruse CA, Cepeda L, Owens B, Johnson SD, Stears J, Lillehei KO. Treatment of recurrent glioma with intracavitary alloreactive cytotoxic T lymphocytes and interleukin-2. Cancer Immunol Immunother 1997;45(2):77–87. https://doi.org/10.1007/s002620050405.
[44] Tsuboi K, Saijo K, Ishikawa E, Tsurushima H, Takano S, Morishita Y, et al. Effects of local injection of ex vivo expanded autologous tumor-specific T lymphocytes in cases with recurrent malignant gliomas. Clin Cancer Res 2003;9(9):3294–302.
[45] Houot R, Schultz LM, Marabelle A, Kohrt H. T-cell-based immunotherapy: adoptive cell transfer and checkpoint inhibition. Cancer Immunol Res 2015;3(10):1115–22. https://doi.org/10.1158/2326-6066.CIR-15-0190.
[46] Rosenberg SA, Packard BS, Aebersold PM, Solomon D, Topalian SL, Toy ST, et al. Use of tumor-infiltrating lymphocytes and interleukin-2 in the immunotherapy of patients with metastatic melanoma. A preliminary report. N Engl J Med 1988;319(25):1676–80. https://doi.org/10.1056/NEJM198812223192527.
[47] Rosenberg SA, Yang JC, Sherry RM, Kammula US, Hughes MS, Phan GQ, et al. Durable complete responses in heavily pretreated patients with metastatic melanoma using T-cell transfer immunotherapy. Clin Cancer Res 2011;17(13):4550–7. https://doi.org/10.1158/1078-0432.CCR-11-0116.
[48] Dudley ME, Wunderlich JR, Yang JC, Sherry RM, Topalian SL, Restifo NP, et al. Adoptive cell transfer therapy following non-myeloablative but lymphodepleting chemotherapy for the treatment of patients with refractory metastatic melanoma. J Clin Oncol 2005;23(10):2346–57. https://doi.org/10.1200/JCO.2005.00.240.
[49] Gattinoni L, Finkelstein SE, Klebanoff CA, Antony PA, Palmer DC, Spiess PJ, et al. Removal of homeostatic cytokine sinks by lymphodepletion enhances the efficacy of adoptively transferred tumor-specific CD8+ T cells. J Exp Med 2005;202(7):907–12. https://doi.org/10.1084/jem.20050732.
[50] Klebanoff CA, Finkelstein SE, Surman DR, Lichtman MK, Gattinoni L, Theoret MR, et al. IL-15 enhances the in vivo antitumor activity of tumor-reactive CD8+ T cells. Proc Natl Acad Sci U S A 2004;101(7):1969–74. https://doi.org/10.1073/pnas.0307298101.
[51] Dudley ME, Wunderlich JR, Shelton TE, Even J, Rosenberg SA. Generation of tumor-infiltrating lymphocyte cultures for use in adoptive transfer therapy for melanoma patients. J Immunother 2003;26(4):332–42. https://doi.org/10.1097/00002371-200307000-00005.
[52] Lin Y, Okada H. Cellular immunotherapy for malignant gliomas. Expert Opin Biol Ther 2016;16(10):1265–75. https://doi.org/10.1080/14712598.2016.1214266.
[53] Berghoff AS, Kiesel B, Widhalm G, Rajky O, Ricken G, Wöhrer A, et al. Programmed death ligand 1 expression and tumor-infiltrating lymphocytes in glioblastoma. Neuro-Oncology 2015;17(8):1064–75. https://doi.org/10.1093/neuonc/nou307.
[54] Maxwell R, Luksik AS, Garzon-Muvdi T, Lim M. The potential of cellular- and viral-based immunotherapies for malignant glioma-dendritic cell vaccines, adoptive cell transfer, and oncolytic viruses. Curr Neurol Neurosci Rep 2017;17(6):50. https://doi.org/10.1007/s11910-017-0754-x.
[55] Wherry EJ, Kurachi M. Molecular and cellular insights into T cell exhaustion. Nat Rev Immunol 2015;15(8):486–99. https://doi.org/10.1038/nri3862.
[56] Quattrocchi KB, Miller CH, Cush S, Bernard SA, Dull ST, Smith M, et al. Pilot study of local autologous tumor infiltrating lymphocytes for the treatment of recurrent malignant gliomas. J Neuro-Oncol 1999;45(2):141–57.
[57] Zhou J, Shen X, Huang J, Hodes RJ, Rosenberg SA, Robbins PF. Telomere length of transferred lymphocytes correlates with in vivo persistence and tumor regression in melanoma patients receiving cell transfer therapy. J Immunol 2005;175(10):7046–52. https://doi.org/10.4049/jimmunol.175.10.7046.
[58] Shen X, Zhou J, Hathcock KS, Robbins P, Powell DJ, Rosenberg SA, et al. Persistence of tumor infiltrating lymphocytes in adoptive immunotherapy correlates with telomere length. J Immunother 2007;30(1):123–9. https://doi.org/10.1097/01.cji.0000211321.07654.b8.

[59] Kaneko S, Mastaglio S, Bondanza A, Ponzoni M, Sanvito F, Aldrighetti L, et al. IL-7 and IL-15 allow the generation of suicide gene-modified alloreactive self-renewing central memory human T lymphocytes. Blood 2009;113(5):1006–15. https://doi.org/10.1182/blood-2008-05-156059.

[60] Golovina TN, Vonderheide RH. Regulatory T cells: overcoming suppression of T-cell immunity. Cancer J 2010;16(4):342–7. https://doi.org/10.1097/PPO.0b013e3181eb336d.

[61] Morita R, Hirohashi Y, Sato N. Depletion of Tregs in vivo: a promising approach to enhance antitumor immunity without autoimmunity. Immunotherapy 2012;4(11):1103–5. https://doi.org/10.2217/imt.12.116.

[62] Robbins PF, Morgan RA, Feldman SA, Yang JC, Sherry RM, Dudley ME, et al. Tumor regression in patients with metastatic synovial cell sarcoma and melanoma using genetically engineered lymphocytes reactive with NY-ESO-1. J Clin Oncol 2011;29(7):917–24. https://doi.org/10.1200/JCO.2010.32.2537.

[63] Rosenberg SA, Kochenderfer JN. Personalized cell transfer immunotherapy for B-cell malignancies and solid cancers. Mol Ther 2011;19(11):1928–30. https://doi.org/10.1038/mt.2011.223.

[64] Sadelain M. T-cell engineering for cancer immunotherapy. Cancer J 2009;15(6):451–5. https://doi.org/10.1097/PPO.0b013e3181c51f37.

[65] Cavalieri S, Cazzaniga S, Geuna M, Magnani Z, Bordignon C, Naldini L, et al. Human T lymphocytes transduced by lentiviral vectors in the absence of TCR activation maintain an intact immune competence. Blood 2003;102(2):497–505. https://doi.org/10.1182/blood-2003-01-0297.

[66] Circosta P, Granziero L, Follenzi A, Vigna E, Stella S, Vallario A, et al. T cell receptor (TCR) gene transfer with lentiviral vectors allows efficient redirection of tumor specificity in naive and memory T cells without prior stimulation of endogenous TCR. Hum Gene Ther 2009;20(12):1576–88. https://doi.org/10.1089/hum.2009.117.

[67] Magnani CF, Turazzi N, Benedicenti F, Calabria A, Tenderini E, Tettamanti S, et al. Immunotherapy of acute leukemia by chimeric antigen receptor-modified lymphocytes using an improved Sleeping Beauty transposon platform. Oncotarget 2016;7(32):51581–97. https://doi.org/10.18632/oncotarget.9955.

[68] Morgan RA, Dudley ME, Wunderlich JR, Hughes MS, Yang JC, Sherry RM, et al. Cancer regression in patients after transfer of genetically engineered lymphocytes. Science 2006;314(5796):126–9. https://doi.org/10.1126/science.1129003.

[69] Bendle GM, Linnemann C, Hooijkaas AI, Bies L, de Witte MA, Jorritsma A, et al. Lethal graft-versus-host disease in mouse models of T cell receptor gene therapy. Nat Med 2010;16(5):565–70. 1p following 70 https://doi.org/10.1038/nm.2128.

[70] Ikeda H, Shiku H. Antigen-receptor gene-modified T cells for treatment of glioma. Adv Exp Med Biol 2012;746:202–15. https://doi.org/10.1007/978-1-4614-3146-6_16.

[71] Abate-Daga D, Davila ML. CAR models: next-generation CAR modifications for enhanced T-cell function. Mol Ther Oncolytics 2016;3:16014. https://doi.org/10.1038/mto.2016.14.

[72] O'Rourke DM, Nasrallah MP, Desai A, Melenhorst JJ, Mansfield K, Morrissette JJD, et al. A single dose of peripherally infused EGFRvIII-directed CAR T cells mediates antigen loss and induces adaptive resistance in patients with recurrent glioblastoma. Sci Transl Med 2017;9(399). https://doi.org/10.1126/scitranslmed.aaa0984.

[73] Bagley SJ, Desai AS, Linette GP, June CH, O'Rourke DM. CAR T-cell therapy for glioblastoma: recent clinical advances and future challenges. Neuro-Oncology 2018;20(11):1429–38. https://doi.org/10.1093/neuonc/noy032.

[74] Buerki RA, Chheda ZS, Okada H. Immunotherapy of primary brain tumors: facts and hopes. Clin Cancer Res 2018;24(21):5198–205. https://doi.org/10.1158/1078-0432.CCR-17-2769.

[75] Brown CE, Warden CD, Starr R, Deng X, Badie B, Yuan YC, et al. Glioma IL13Ralpha2 is associated with mesenchymal signature gene expression and poor patient prognosis. PLoS ONE 2013;8(10). https://doi.org/10.1371/journal.pone.0077769, e77769.

[76] Debinski W, Gibo DM, Hulet SW, Connor JR, Gillespie GY. Receptor for interleukin 13 is a marker and therapeutic target for human high-grade gliomas. Clin Cancer Res 1999;5(5):985–90.

[77] Thaci B, Brown CE, Binello E, Werbaneth K, Sampath P, Sengupta S. Significance of interleukin-13 receptor alpha 2-targeted glioblastoma therapy. Neuro-Oncology 2014;16(10):1304–12. https://doi.org/10.1093/neuonc/nou045.

[78] Wykosky J, Gibo DM, Stanton C, Debinski W. Interleukin-13 receptor alpha 2, EphA2, and Fos-related antigen 1 as molecular denominators of high-grade astrocytomas and specific targets for combinatorial therapy. Clin Cancer Res 2008;14(1):199–208. https://doi.org/10.1158/1078-0432.ccr-07-1990.

[79] Brown CE, Badie B, Barish ME, Weng L, Ostberg JR, Chang WC, et al. Bioactivity and safety of IL13Rα2-redirected chimeric antigen receptor CD8+ T cells in patients with recurrent glioblastoma. Clin Cancer Res 2015;21(18):4062–72. https://doi.org/10.1158/1078-0432.CCR-15-0428.

[80] Wang RF. The role of MHC class II-restricted tumor antigens and CD4+ T cells in antitumor immunity. Trends Immunol 2001;22(5):269–76. https://doi.org/10.1016/s1471-4906(01)01896-8.
[81] Tsurushima H, Liu SQ, Tsuboi K, Yoshii Y, Nose T, Ohno T. Induction of human autologous cytotoxic T lymphocytes against minced tissues of glioblastoma multiforme. J Neurosurg 1996;84(2):258–63. https://doi.org/10.3171/jns.1996.84.2.0258.
[82] Tsurushima H, Liu SQ, Tuboi K, Matsumura A, Yoshii Y, Nose T, et al. Reduction of end-stage malignant glioma by injection with autologous cytotoxic T lymphocytes. Jpn J Cancer Res 1999;90(5):536–45. https://doi.org/10.1111/j.1349-7006.1999.tb00781.x.
[83] Holladay FP, Heitz-Turner T, Bayer WL, Wood GW. Autologous tumor cell vaccination combined with adoptive cellular immunotherapy in patients with grade III/IV astrocytoma. J Neuro-Oncol 1996;27(2):179–89. https://doi.org/10.1007/bf00177482.
[84] Plautz GE, Miller DW, Barnett GH, Stevens GH, Maffett S, Kim J, et al. T cell adoptive immunotherapy of newly diagnosed gliomas. Clin Cancer Res 2000;6(6):2209–18.
[85] Sloan AE, Dansey R, Zamorano L, Barger G, Hamm C, Diaz F, et al. Adoptive immunotherapy in patients with recurrent malignant glioma: preliminary results of using autologous whole-tumor vaccine plus granulocyte-macrophage colony-stimulating factor and adoptive transfer of anti-CD3-activated lymphocytes. Neurosurg Focus 2000;9(6), e9.
[86] Wood GW, Holladay FP, Turner T, Wang YY, Chiga M. A pilot study of autologous cancer cell vaccination and cellular immunotherapy using anti-CD3 stimulated lymphocytes in patients with recurrent grade III/IV astrocytoma. J Neuro-Oncol 2000;48(2):113–20.
[87] Nair SK, De Leon G, Boczkowski D, Schmittling R, Xie W, Staats J, et al. Recognition and killing of autologous, primary glioblastoma tumor cells by human cytomegalovirus pp65-specific cytotoxic T cells. Clin Cancer Res 2014;20(10):2684–94. https://doi.org/10.1158/1078-0432.CCR-13-3268.
[88] Nair SK, Sampson JH, Mitchell DA. Immunological targeting of cytomegalovirus for glioblastoma therapy. Onco Targets Ther 2014;3. https://doi.org/10.4161/onci.29289, e29289.
[89] Nair DLG, Boczkowski D, Schmittling R, Xie W, Archer G, et al. Targeting cytomegalovirus (CMV) antigens for GBM immunotherapy. J Immunother Cancer 2013;1(Suppl 1):P270. https://doi.org/10.1186/2051-1426-1-s1-p270.
[90] Schuessler A, Smith C, Beagley L, Boyle GM, Rehan S, Matthews K, et al. Autologous T-cell therapy for cytomegalovirus as a consolidative treatment for recurrent glioblastoma. Cancer Res 2014;74(13):3466–76. https://doi.org/10.1158/0008-5472.can-14-0296.
[91] Flores C, Pham C, Snyder D, Yang S, Sanchez-Perez L, Sayour E, et al. Novel role of hematopoietic stem cells in immunologic rejection of malignant gliomas. Onco Targets Ther 2015;4(3). https://doi.org/10.4161/2162402X.2014.994374, e994374.
[92] Wildes TJ, Grippin A, Dyson KA, Wummer BM, Damiani DJ, Abraham RS, et al. Cross-talk between T cells and hematopoietic stem cells during adoptive cellular therapy for malignant glioma. Clin Cancer Res 2018. https://doi.org/10.1158/1078-0432.CCR-17-3061.
[93] Flores CT, Wildes TJ, Drake JA, Moore GL, Dean BD, Abraham RS, et al. $Lin^{-}CCR2^{+}$ hematopoietic stem and progenitor cells overcome resistance to PD-1 blockade. Nat Commun 2018;9(1):4313. https://doi.org/10.1038/s41467-018-06182-5.
[94] Flores C, Wildes T, Dean BD, Moore G, Drake J, Abraham R, et al. Massive clonal expansion of medulloblastoma-specific T cells during adoptive cellular therapy. Sci Adv 2019;5(11). https://doi.org/10.1126/sciadv.aav9879, eaav9879.
[95] Sridhar P, Petrocca F. Regional delivery of chimeric antigen receptor (CAR) T-cells for cancer therapy. Cancers (Basel) 2017;9(7). https://doi.org/10.3390/cancers9070092.
[96] Gust J, Hay KA, Hanafi LA, Li D, Myerson D, Gonzalez-Cuyar LF, et al. Endothelial activation and blood-brain barrier disruption in neurotoxicity after adoptive immunotherapy with CD19 CAR-T cells. Cancer Discov 2017;7(12):1404–19. https://doi.org/10.1158/2159-8290.CD-17-0698.
[97] Wang Z, Han W. Biomarkers of cytokine release syndrome and neurotoxicity related to CAR-T cell therapy. Biomark Res 2018;6:4. https://doi.org/10.1186/s40364-018-0116-0.
[98] Taraseviciute A, Tkachev V, Ponce R, Turtle CJ, Snyder JM, Liggitt HD, et al. Chimeric antigen receptor T cell-mediated neurotoxicity in nonhuman primates. Cancer Discov 2018;8(6):750–63. https://doi.org/10.1158/2159-8290.CD-17-1368.

[99] Gardner R, Leger KJ, Annesley CE, Summers C, Rivers J, Gust J, et al. Decreased rates of severe CRS seen with early intervention strategies for CD19 CAR-T cell toxicity management. Blood 2016;128(22):586. https://doi.org/10.1182/blood.V128.22.586.586.

[100] Choi BD, Suryadevara CM, Gedeon PC, Herndon JE, Sanchez-Perez L, Bigner DD, et al. Intracerebral delivery of a third generation EGFRvIII-specific chimeric antigen receptor is efficacious against human glioma. J Clin Neurosci 2014;21(1):189–90. https://doi.org/10.1016/j.jocn.2013.03.012.

[101] June CH, O'Connor RS, Kawalekar OU, Ghassemi S, Milone MC. CAR T cell immunotherapy for human cancer. Science 2018;359(6382):1361–5. https://doi.org/10.1126/science.aar6711.

[102] Keu KV, Witney TH, Yaghoubi S, Rosenberg J, Kurien A, Magnusson R, et al. Reporter gene imaging of targeted T cell immunotherapy in recurrent glioma. Sci Transl Med 2017;9(373). https://doi.org/10.1126/scitranslmed.aag2196.

[103] Ronald JA, Kim BS, Gowrishankar G, Namavari M, Alam IS, D'Souza A, et al. A PET imaging strategy to visualize activated T cells in acute graft-versus-host disease elicited by allogenic hematopoietic cell transplant. Cancer Res 2017;77(11):2893–902. https://doi.org/10.1158/0008-5472.CAN-16-2953.

[104] Alam IS, Mayer AT, Sagiv-Barfi I, Wang K, Vermesh O, Czerwinski DK, et al. Imaging activated T cells predicts response to cancer vaccines. J Clin Invest 2018;128(6):2569–80. https://doi.org/10.1172/JCI98509.

[105] Gonzales C, Yoshihara HA, Dilek N, Leignadier J, Irving M, Mieville P, et al. In-vivo detection and tracking of T cells in various organs in a melanoma tumor model by 19F-fluorine MRS/MRI. PLoS ONE 2016;11(10). https://doi.org/10.1371/journal.pone.0164557, e0164557.

[106] Liu L, Ye Q, Wu Y, Hsieh WY, Chen CL, Shen HH, et al. Tracking T-cells in vivo with a new nano-sized MRI contrast agent. Nanomedicine 2012;8(8):1345–54. https://doi.org/10.1016/j.nano.2012.02.017.

[107] Keskin DB, Anandappa AJ, Sun J, Tirosh I, Mathewson ND, Li S, et al. Neoantigen vaccine generates intratumoral T cell responses in phase Ib glioblastoma trial. Nature 2019;565(7738):234–9. https://doi.org/10.1038/s41586-018-0792-9.

CHAPTER

6

Immune checkpoint blockade therapy in high-grade glioma

Jessica Waibl Polania[a,b], *Selena Lorrey*[a,c], *Daniel Wilkinson*[a], *and Peter E. Fecci*[a,b,c]

[a]Duke Brain Tumor Immunotherapy Program, Department of Neurosurgery, Duke University Medical Center, Durham, NC, United States [b]Department of Pathology, Duke University Medical Center, Durham, NC, United States [c]Department of Immunology, Duke University Graduate School, Durham, NC, United States

Abstract

Glioblastoma (GBM) is the most common primary brain cancer in adults and remains universally lethal. Despite extensive medical intervention, the median survival remains a bleak 15–17 months from time of diagnosis, calling for new therapeutic strategies. In recent years, immune checkpoint blockade (ICB) has garnered attention for its clinical successes in extending overall survival in many advanced-stage cancers. To date, ICB clinical trials have failed to proffer the same benefit in GBM patients, turning attention in the field toward understanding this shortcoming and employing new strategies. Genomic characteristics such as mutational burden, immunogenicity, and heterogeneity, along with immune infiltrate, CNS location, and immunosuppression (tumor-induced or iatrogenic), have been identified as critical components of ICB responsiveness. Promising results from recent clinical trials have rewarded efforts to understand the features of successful ICB and to restore enthusiasm for ICB as an encouraging therapeutic in the treatment of GBM.

Abbreviations

APC	antigen-presenting cell
BBB	blood-brain barrier
CNS	central nervous system
CSF	cerebrospinal fluid
CTLA-4	cytotoxic T lymphocyte antigen 4
FDA	Food and Drug Administration
GBM	glioblastoma
ICB	immune checkpoint blockade
LAG-3	lymphocyte activating 3
MHC	major histocompatibility complex

https://doi.org/10.1016/B978-0-12-819755-4.00002-3

MMR	mismatch repair
MMRD	mismatch repair deficient
MSI	microsatellite instability
NSCLC	nonsmall-cell lung cancer
ORR	objective response rates
OS	overall survival
PD-1	programmed cell death 1
PD-L1	programmed death ligand 1
PD-L2	programmed death ligand 2
TAM	tumor-associated macrophage
TCR	T-cell receptor
TIGIT	T-cell immunoreceptor with Ig and ITIM domains
TIM-3	T-cell immunoglobulin and mucin-domain containing-3
TMB	tumor mutational burden
TME	tumor microenvironment
TMZ	temozolomide
T_{reg}	regulatory T cell
VEGF	vascular endothelial growth factor

Conflict of interest

No potential conflicts of interest were disclosed.

Introduction

Glioblastoma (GBM) is the most common primary brain cancer in adults and remains universally lethal. Despite extensive medical intervention, the median survival remains a bleak 15–17 months from time of diagnosis [1–3]. Plagued with abnormally high cellular heterogeneity, robust immunosuppression, and near 100% recurrence, GBM has been especially resistant to the standard treatment of surgical resection, chemotherapy, and radiation. In recent years, immune checkpoint blockade (ICB) has emerged as an effective treatment for several advanced-stage cancers. The success of ICB began in 2010 with a landmark phase III randomized clinical trial in advanced melanoma patients that led to US Food and Drug Administration (FDA) approval of ipilimumab (anti-CTLA-4, Bristol-Myers-Squibb). Approval was secondary to improved survival seen in patients treated with ipilimumab alone or in combination with a gp100 vaccine [4]. Not long after, early trials of anti-PD-1 treatment successfully promoted durable antitumor responses, which led to the FDA approval of nivolumab (anti-PD-1, Bristol-Meyers-Squibb) and pembrolizumab (anti-PD-1, Merck), both for advanced melanoma [5]. In 2018, the Nobel Prize was awarded to James P. Allison, PhD, of The University of Texas MD Anderson Cancer Center, and Dr. Tasuku Honjo of Kyoto University in Japan, for their research on CTLA-4 and PD-1, respectively.

Excitement surrounding ICB has led to many clinical trials testing its efficacy in a variety of cancers, including GBM [3, 4, 6–8]. Results in GBM have been mostly unsatisfactory, with the first randomized, phase III trial of nivolumab for recurrent GBM ending prematurely after failure to improve overall survival over standard-of-care plus bevacizumab treatment [9]. This failure, among others, has shifted the focus of the field toward developing a deeper understanding of how and when ICB fails or succeeds [5]. In this chapter, we explain the homeostatic roles played by immune checkpoint molecules, describe hallmarks of successful ICB, and discuss the intrinsic and extrinsic features of GBM which have limited success in this cancer thus far. We also summarize pertinent clinical trials and promising new ICB strategies.

Coinhibitory molecules: A role in homeostasis

Immune checkpoint blockade has gained renown for its success in several cancer types in recent years. However, "immune checkpoints" do not exist solely as a target for immunotherapy; instead, these coinhibitory molecules are responsible for dampening T-cell activation under physiologic conditions. It is the balance between coinhibitory and costimulatory signaling that guides successful T-cell activation for a controlled period of time and determines the depth and duration of the T-cell response.

At the cellular level, T-cell activation requires three signals: interaction between the T-cell receptor (TCR) and peptide-loaded major histocompatibility complex (MHC) on antigen-presenting cells (APCs) (peptide:MHC complex); costimulation; and cytokine signals [10]. In contrast to costimulatory molecules, coinhibitory molecules serve to moderate the T-cell response to a given stimulus. Too small a response results in inadequate pathogen clearance, while too large a response can result in unbridled inflammation and collateral tissue damage, trending toward autoimmunity.

The regulation of T-cell activity is described well by the "Tide Model" of cosignaling put forth by Yuwen Zhu et al., which likens T-cell activation to incoming, peak, and outgoing tides [11]. The "incoming tide" describes the early phase when naïve T cells are just becoming activated. Both the naïve and the newly activated T cells express high levels of costimulatory molecules on their cell surface, with fewer coinhibitory molecules. At "peak tide," there are now many activated T cells responding to a particular stimulus. Both costimulatory and coinhibitory molecules are expressed, and these oppose one another to simply maintain the responding T-cell population. In contrast, the "outgoing tide" is the late period when the pathogen or stimulus has been cleared, and there is no longer a need for activated T cells. During this stage, coinhibitory molecules are predominantly present on the T-cell surface to suppress T-cell activity, thereby reinforcing peripheral tolerance and preventing autoimmunity [11].

This balance (or imbalance) of costimulatory and coinhibitory molecules may also engender the phases of cancer immunoediting. According to the cancer immunoediting hypothesis, tumors may progress through phases of immunologic elimination, equilibrium, and eventual escape. The particular phase embodied at any moment may reflect, in part, the cosignaling balance within T cells [12] (see Fig. 1).

Likewise, a lack of T-cell costimulatory molecules and the concomitant presence of coinhibitory molecules can contribute to immunosuppression in cancer. The most well-recognized costimulatory molecule is CD28, which is constitutively expressed on resting T cells and is the primary costimulatory molecule for naïve T-cell priming. The canonical "signal 2" represents the binding of CD28 to a B7 molecule (one of two functionally equivalent molecules, B7-1 or B7-2, otherwise known as CD80 and CD86, respectively) on an APC [13]. The most well-recognized coinhibitory molecules, in turn, are cytotoxic T lymphocyte antigen-4 (CTLA-4) and programmed cell death protein-1 (PD-1).

CTLA-4 is a homolog of CD28 that binds B7 (either molecule) with 10- to 20-fold higher affinity than CD28. While CD28 is constitutively expressed on T cells, CTLA-4 is inducible,

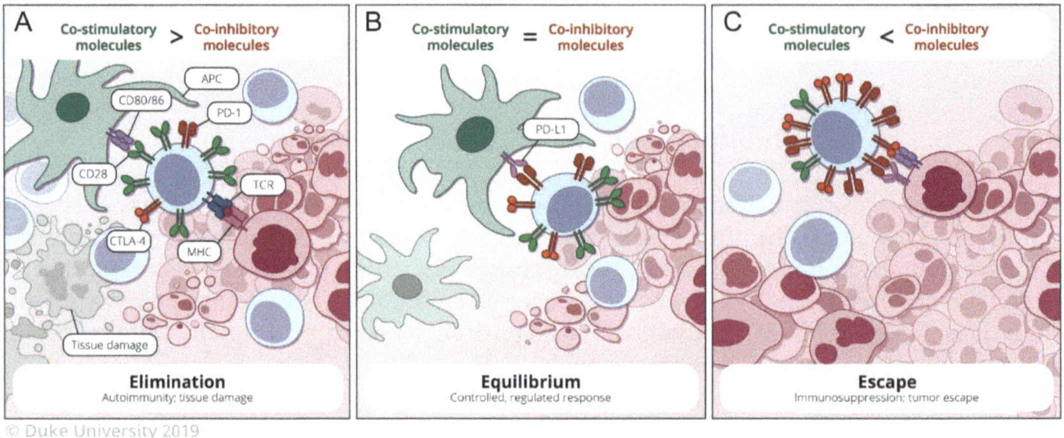

FIG. 1 The role of cosignaling molecules in T-cell activation in response to tumor antigens. (A) A wealth of costimulatory molecules on tumor cells or immune cells in the microenvironment can facilitate tumor elimination, albeit with the risk of uncontrolled T-cell activation and subsequent autoimmunity and/or tissue damage. (B) A balance of coinhibitory and costimulatory molecules on T cells or tumor cells may permit an equilibrium to develop between the tumor and the immune response, leading to tumor persistence. (C) Imbalance toward coinhibitory molecules on tumor cells or immune cells may instigate tumor escape. *Illustrated by Megan Llewellyn, MSMI, CMI; copyright Duke University; with permission under a CC BY-ND 4.0 license.*

and cell surface expression is upregulated after signals from both the TCR and CD28 [14, 15]. By outcompeting CD28 for B7 binding, and thereby thwarting an activating signal, CTLA-4 functions to maintain peripheral tolerance and limit T-cell activation upon its induction. Although inducible on conventional T cells, CTLA-4 is constitutively expressed at high levels on T_{regs} and contributes to T_{reg}-mediated immune suppression [14].

PD-1 is a coinhibitory molecule expressed on the surface of many cell types, including activated T cells. Binding of PD-1 to its ligand PD-L1 or PD-L2 results in the inhibition of TCR signaling and subsequent T-cell effector functions. Both PD-L1 and PD-L2 are often upregulated on various cancers, including GBM. PD-L1 is constitutively expressed on most tissues and is upregulated on a number of cells in the tumor microenvironment, including activated T cells, B cells, monocytes, DCs, and vascular endothelial cells in response to IFN-γ signaling [16–18]. In contrast, PD-L2 is expressed mainly on APCs and has a fourfold higher affinity for PD-1, as compared with PD-L1 [18].

In addition to the canonical checkpoints PD-1 and CTLA-4, there are other coinhibitory molecules that serve similar functions in T-cell homeostasis and are potential targets for immunotherapy. These include lymphocyte activation gene 3 protein (LAG-3), T-cell immunoglobulin and mucin domain 3 protein (TIM-3), and the T-cell immunoreceptor with Ig and ITIM domains protein (TIGIT). LAG-3 shares homology with CD4, and as such, binds MHCII, but with a higher affinity than CD4 [19]. LAG-3 is found on both CD4 and CD8 T cells as well as NK cells [20]. TIM-3 is found on DCs, monocytes, and NK cells, as well as T cells. Lastly, TIGIT is an inducible coinhibitory molecule expressed on T cells, most highly on T_{regs} and Tfh cells, and, like CTLA-4, competes with higher affinity for its ligand CD155 or CD112 on APCs than its costimulatory counterpart CD266 [20, 21].

Tumor-induced mechanisms of immunosuppression and T-cell dysfunction have been reviewed elsewhere [22], but it is important to note that coinhibitory molecules play a role in both exhaustion and tolerance. Exhausted T cells can express multiple coinhibitory receptors, including PD-1, CTLA-4, LAG-3, CTLA4, TIGIT, and TIM3 [23, 24]. Further studies are necessary to understand how coinhibitory receptors regulate T-cell activities and whether they functionally contribute to exhaustion or not. While many coinhibitory receptors are exciting targets for cancer immunotherapy, particularly in combination, one must understand and consider the potentially complementary functions of these molecules. Treatment in GBM with checkpoint inhibitor monotherapy has proven ineffective, and studies have naturally progressed to targeting two or more checkpoints. The goal for treatment is to tip the scales in favor of costimulatory pathways without risking autoimmunity within the central nervous system (CNS). Deciding on appropriate combinations of immune checkpoint targets requires an advanced knowledge of the specific functions for each checkpoint.

Preclinical and clinical studies of ICB

Preclinical success with immune checkpoint blockade

Several checkpoint blockade therapies targeting CTLA-4 and PD-1/PD-L1 have recently gained FDA approval for multiple cancers. Early studies in mouse tumor models initially targeted the immune checkpoint CTLA-4. In 1996, Leach, et al. showed that anti-CTLA-4 enhanced the rejection of a subcutaneously implanted murine colon carcinoma line transfected with B7-1 [25]. Notably, anti-CTLA-4 also inhibited the growth of a tumor cell line lacking B7 expression, suggesting a mechanistic component independent of T cell: tumor interactions. CTLA-4 blockade also prevented tumor growth upon rechallenge and was effective against established tumors. Subsequent studies demonstrated success in mouse models of prostate [26], colon [27], and breast [28] carcinoma, as well as in models of thymoma [29] and melanoma [30]. Regarding PD-1 blockage, preclinical studies illustrated the benefit of targeting the PD-1/PD-L1 axis in B16 melanoma and CT26 colon carcinoma models [31], as well as additional murine tumor types [32–34].

Clinical trials of immune checkpoint blockade in nongliomas

Promising preclinical results led to clinical trials exploring the efficacy of ICB in various cancers. Hodi et al. administered the anti-CTLA-4 antibody MDX-CTLA-4 (ipilimumab) to nine patients with metastatic melanoma or ovarian carcinoma [35]. There were no high-grade toxicities associated with the antibody, and several patients showed increased antitumor immunity following treatment, paving the way for future anti-CTLA-4 trials. A phase III trial in which ipilimumab was administered to advanced melanoma patients resulted in a 3.5-month survival benefit compared with the control arm [4]. This resulted in FDA approval of ipilimumab for patients with metastatic melanoma in 2011.

The first reported use of anti-PD-1 in clinical trials was published in 2010 [36]. In this phase I study, a fully human anti-PD-1 (MDX-1106; nivolumab) administered to patients with various solid tumors was deemed relatively well tolerated and safe. Subsequent trials showed considerable objective response rates (ORRs) in patients with nonsmall-cell lung carcinoma (NSCLC), melanoma, and renal cell carcinoma [37]. Importantly, higher response rates correlated with higher expression of PD-L1 on tumor cells. Nivolumab and pembrolizumab, both anti-PD-1 antibodies, received FDA approval for advanced melanoma in 2014. There is an ever-increasing list of cancers for which CTLA-4 and PD-1 blockade are FDA-approved, with preclinical and clinical studies underway in numerous additional tumor types, including glioma.

Preclinical studies of immune checkpoint blockade in glioma

Human and murine glioma-infiltrating T cells express high levels of checkpoints and prove to be severely exhausted [38], making ICB a potentially promising therapeutic modality for brain tumors. ICB has been tested preclinically in several immunocompetent murine glioma models, with varying results. In an early study, systemic anti-CTLA-4 resulted in 80% long-term survival (>100 days) among mice bearing SMA-560 glioma, with no evidence of experimental autoimmune encephalomyelitis (EAE) [39]. In another study, anti-CTLA-4, -PD-1, -PD-L1, or -PD-L2 monotherapy, or a combination of CTLA-4 and PD-1 blockade, was tested in the GL261 mouse model of glioma [40]. Combination therapy resulted in 75% long-term survival, compared with 15% and 50% for anti-CTLA-4 or anti-PD-1 monotherapy, respectively. Anti-PD-L1 led to 25% long-term survival, while anti-PD-L2 did not have an effect in this study. An additional study in GL261 glioma resulted in 60% long-term survival in mice treated with anti-PD-L1 [41]. Interestingly, a separate study showed an increased efficacy for anti-PD-1 in GL261 glioma when blockade was combined with radiation [42]. A recently derived mouse glioma cell line, SB28, may accurately recapitulate the low immunogenicity of human GBM [43]. Anti-PD-1 therapy had little to no effect on survival in mice bearing SB28 tumors [44]. Together, these studies have formed the basis for clinical trials with ICB in GBM that are already completed or now underway.

Clinical trials of immune checkpoint blockade in GBM

Although there are a number of ongoing and completed clinical trials testing the efficacy of ICB in glioma, there has been very little success. The largest trial to date is Checkmate 143 (NCT02017717), originally designed to examine the safety and tolerability of nivolumab (anti-PD-1) as a monotherapy or in combination with ipilimumab (anti-CTLA-4) in recurrent GBM. There were no grade 3–4 adverse effects in patients receiving nivolumab monotherapy, but 30%–90% of patients receiving the combination of nivolumab and ipilimumab had grade 3–4 adverse events, depending on the dose [45]. Stable disease, or better, was observed in 6/10, 4/10, and 9/20 patients receiving 3 mg/kg nivolumab every 2 weeks, nivolumab 1 mg/kg + ipilimumab 3 mg/kg every 3 weeks, and nivolumab 3 mg/kg + ipilimumab 1 mg/kg for 4 doses, respectively. In a phase III cohort of Checkmate 143, patients received either nivolumab or bevacizumab, an anti-VEGF antibody. Nivolumab did not confer the improved median overall survival compared with bevacizumab (9.8 months vs 10 months, respectively) [9]. However, nivolumab had a longer response duration than bevacizumab (11.1 months and 5.3 months, respectively). In a retrospective case series, nivolumab efficacy was evaluated in recurrent GBM patients that progressed following bevacizumab therapy [46]. Nivolumab failed to prolong survival in these bevacizumab-refractory patients.

Clinical trials with pembrolizumab (anti-PD-1) are also underway in gliomas. In Keynote-028, 25 patients with recurrent GBM were administered a standard dose of pembrolizumab (10 mg/kg every 2 weeks), with the primary outcomes being tolerability and overall response [47]. Importantly, inclusion criteria included PD-L1 expression on greater than 1% of tumor or stroma cells. One patient experienced a partial response (4% ORR), and 12 patients showed stable disease (48% SD). In a single institution study, pembrolizumab was administered to 25 patients with high-grade gliomas [48]. Out of 24 patients whose outcomes were analyzed, 2 showed a partial response, 5 showed stable disease, and 17 showed progressive disease. Despite low response rates, several patients experienced prolonged progression-free survival. Additional studies failed to show a benefit of pembrolizumab in recurrent GBM patients [49, 50].

Durvalumab, a fully human monoclonal antibody against PD-L1, is an FDA-approved therapy for a subset of patients with advanced or metastatic bladder cancer, and clinical trials are underway in several other malignancies. A phase II trial with durvalumab in GBM (NCT02336165) has shown that the therapy is well tolerated. Additionally, there was a durable response in a subset of patients that had not received bevacizumab, whereby 4 patients out of 30 remained progression-free after 1 year [51]. A separate cohort from the same trial is exploring the combination of durvalumab with standard radiotherapy in newly diagnosed, unmethylated GBM [52]. As of November 2018, overall survival at 12 months was 60% in treated patients, compared with 50% at 12 months in historical controls. Median overall survival was 15.1 months in treated patients compared with 12.7 months in historical controls. Clinical trials testing additional anti-PD-L1 (atezolizumab) and anti-PD-1 (pidilizumab) antibodies are in early stages, demonstrating that although results have been underwhelming thus far, there is still excitement surrounding the use of ICB therapy in GBM. Understanding ICB failures in GBM assumes critical importance as we look to adjust strategies with these therapies (see "New approaches/future directions" section below).

Why has ICB struggled in GBM?

To better understand the limited efficacy of ICB in GBM, it is important to recognize the current components to successful ICB in other cancers. The first requirement is an ample and functional T-cell pool capable of being primed and trafficking to tumor. Likewise, a cross-priming APC population is needed. Once primed T cells reach the tumor, their TCRs must engage peptide-MHC complexes on the tumor, where appropriate costimulatory signaling will lead to perforin-mediated tumor cell lysis [53, 54]. Additionally, T cells must be able to maintain their function in the tumor microenvironment. Tumor intrinsic factors such as mutational burden, immunogenicity, and heterogeneity, along with tumor extrinsic factors such as location, privilege, and immunosuppression (tumor-induced or iatrogenic), are the critical components of the antitumor response and ICB responsiveness (Fig. 2).

Tumor mutational burden/immunogenicity

Most commonly, tumor immune responses arise from the recognition of tumor-specific neoantigens as a result of nonsynonymous somatic mutations in cancer genomes [55–57]. This is best observed in cases of melanoma [58] and NSCLC [59], which have high numbers of

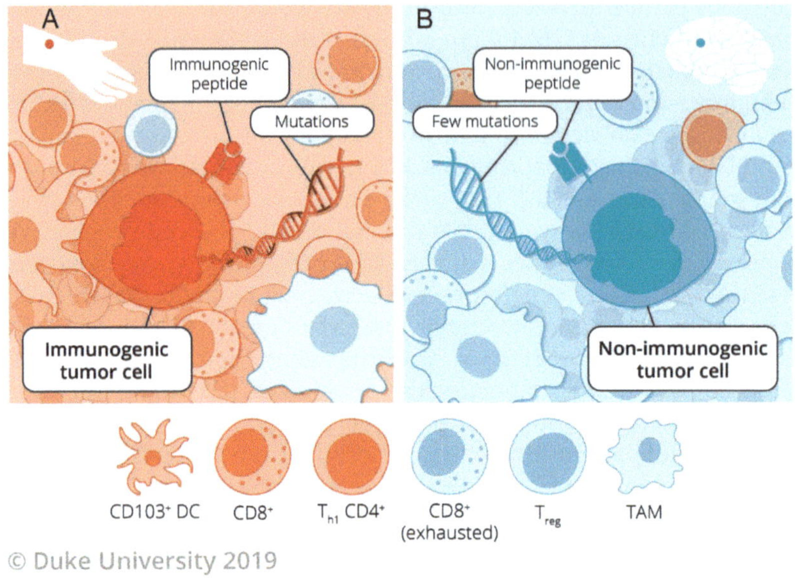

FIG. 2 Cellular and molecular correlates of ICB responsiveness. (A) ICB-responsive tumors, such as melanoma, are characterized by high tumor mutational burden with many immunogenic antigens. These "hot" tumors are often highly infiltrated by Th1 CD4+ T cells, CD8+ T cells, CD103+ DCs, and fewer numbers of immunosuppressive cells such as TAMs and T_{regs}. (B) In contrast, GBMs have relatively few mutations and demonstrate poor immunogenicity. The TME of GBM is dominated by immunosuppressive TAMs and T_{regs}, with few CD8+ T cells. *Illustrated by Megan Llewellyn, MSMI, CMI; copyright Duke University; with permission under a CC BY-ND 4.0 license.*

mutations, referred to as tumor mutational burden (TMB), owing to carcinogenic origins [60]. Higher TMB increases the probability of generating immunogenic neoantigens that lead to robust T-cell responses [61, 62]. In the case of lung cancer, high TMB has been attributed high predictive value for successful ICB therapy [6, 63].

Expectedly, tumor genomic instability, reflected in either high microsatellite instability (MSI) or defects in DNA mismatch repair (MMR) machinery, has correlated with improved disease outcome when such tumors are treated with ICB. This is presumably due to the increased prevalence of nonsynonymous mutations and, consequently, increased immunogenic neoantigens [7, 64]. In 2017, the FDA approved pembrolizumab (anti-PD-1) for the treatment of MSI-high and MMR-defective tumors irrespective of location [7]. This landmark approval is telling of the impact TMB has on the efficacy of ICB therapy. Notably, the quality of tumor neoantigens, i.e., their degree of dissimilarity from "self" or similarity to infectious disease-derived peptides, has proven more important than their quantity in a study of long-term survivors of pancreatic ductal adenocarcinoma [21, 22]. In one study, a neoantigen fitness model ascribing immunogenicity to homology with microbial or viral peptides was able to identify long-term survivors in two independent datasets [21]. Therefore, both neoantigen homology to pathogenic peptides and high neoantigenic burden are crucial to mounting a successful tumor immune response following ICB.

In contrast to ICB-responsive cancers, gliomas harbor a low-average TMB [64]. A retrospective analysis of TMB in 198 GBM samples found 7/198 (3.8%) to have a high TMB (>20 mutations/1.4Mb), while the majority (86.1%) had a low TMB (<10 mutations/1.4Mb) [65]. An exception is MMR-deficient (MMRD) GBM, which is prone to a hypermutator phenotype [66, 67]. A recent study using exome sequencing and neoantigen prediction compared TMB and neoantigen quantities between MMRD GBM and ICB-responsive adult cancers from several tissues. Researchers found that MMRD GBM had a neoantigen load 6–17 times higher than melanoma, lung cancer, and MSI-high GI cancers. On this basis, two patients with bMMRD were treated with nivolumab (anti-PD-1), which resulted in clinically significant responses [66].

In addition to low TMB, extensive cellular and molecular intratumoral heterogeneity has been associated with therapeutic resistance to ICB and recurrence in GBM (reviewed in [68]). A testament to the degree of heterogeneity within GBM is a study using single-cell RNA sequencing that showed a single tumor contained a mixture of different GBM molecular subgroups [69]. Heterogeneity can also be dynamic. Johnson et al. found that in 43% of profiled GBMs, at least half of the mutations found at diagnosis were missing at recurrence [70]. Additionally, single cell-derived subclones can harbor unique genetic identities that confer drug resistance [71]. Subclonal populations not targeted by initial treatment result in expanding treatment resistance and tumor recurrence that can be an obstacle for ICB efficacy.

Immune infiltrate

Several attributes of a tumor's immune infiltrate can affect the efficacy of ICB, including composition, localization, and functionality [72]. In terms of composition, higher numbers of infiltrating CD8 T cells, Th1 CD4 T cells, and CD103 dendritic cells are positively correlated with ICB response, while high amounts of T_{reg} and immunosuppressive tumor-associated macrophages (TAMs) have been linked to the limited efficacy for ICB and other immunotherapies [73, 74]. Localization refers simply to whether antitumor immune cells can access and infiltrate the tumor, which is then categorized as either "hot" or "cold" [75]. Inflamed, "hot" tumors exhibit high levels of CD8 T-cell interaction with neoplastic cells and respond more favorably to ICB compared with deserted, "cold" tumors with low levels of CD8 T cells intratumorally [74]. Finally, how functional an immune infiltrate is able to remain within the microenvironment influences the likelihood for clinical responses [72]. Unsurprisingly, melanoma is among those malignancies harboring an activated antitumor T-cell infiltrate, contributing to its responsiveness to ICB therapy [76, 77].

With regard to the composition, localization, and functionality of immune infiltrates, GBM is on the poor end of the spectrum for all three. Dominated by high levels of immunosuppressive TAMs and T_{regs}, the GBM tumor microenvironment is exceptionally unequipped for an effective T-cell response to ICB therapy [22, 78–80]. Likewise, GBM itself is a notoriously immunosuppressive tumor, constantly outmaneuvering the cellular immune response [79]. The gamut of T-cell dysfunction has been observed locally amidst GBM, including anergy [81, 82], T_{reg}-induced tolerance [78, 83–85], and exhaustion [24, 38]. T-cell exhaustion, a hyporesponsive state resulting from chronic antigen exposure, is targeted for reversal by anti-PD-1 strategies [86, 87]. As described above, PD-1 is expressed by activated T cells and is designed to inhibit T-cell effector function when bound to its ligand PD-L1 or PD-L2 (which are frequently found on GBM cells) [87]. Blocking this interaction can perpetuate T-cell activity and lead to tumor cell lysis [87]. This only works, however, if T cells are in an early state of exhaustion [88]. In the case of GBM, T cells are frequently found in a terminally exhausted state characterized by the upregulation of alternative immune checkpoints (TIM-3, LAG-3, 2B4, TIGIT). Terminal exhaustion is not reversible by ICB alone [22, 23, 38]. In an effort to address this, clinical trials are underway to evaluate the efficacy of TIM-3 and LAG-3 blockade in combination with anti-PD-1 (NCT02658981, NCT02817633).

Additional factors

Intracranial location

In GBM, an additional obstacle to successful antitumor immunity is the tumor's intracranial locale. For decades, the brain was considered an immune-privileged site due to absent rejection of tissue grafts implanted in the CNS parenchyma [89]. The blood-brain barrier (BBB) and the absence of lymphatic vessels were often cited in explanation [90]. Recently, discovery of the glial-lymphatic (glymphatic) system has provided a mechanism for cerebrospinal fluid (CSF)-interstitial fluid exchange and subsequent drainage to deep cervical lymph nodes, as well as a path for antigen and immune cell trafficking to and from the CNS [91–93]. Among other findings over the last 20 years, this has gradually brought a demise to the notion of the brain as a truly privileged site [94]. The BBB, however, does still warrant consideration when designing therapeutic approaches within the CNS.

Classically described as a "barricade to the CNS," the BBB is composed of specialized endothelial cells with tight junctions coated with a fusion of endothelial and astrocyte glia limitans basement membranes. Fluid and low molecular weight molecules pass freely through the BBB but T-cell trafficking is limited, albeit possible through postcapillary venules [90]. This barrier is somewhat overcome in the setting of GBM where the BBB becomes damaged, limiting its restrictive capacity [95, 96]. Taken together, the brain is not as restricted from immunity as previously thought; however, it retains a unique set of challenges which must be taken into account for immunotherapeutic targeting.

Systemic immunosuppression

In addition to local immunosuppression within the tumor environment, GBM patients themselves demonstrate marked defects in systemic immunity. Some of these defects result from tumor-imposed immune subterfuge, while some are the unfortunate result of standard-of-care therapy. Regarding the former, GBM often elicits severe lymphopenia, with many patients displaying decreased levels of circulating T cells, as well as surprising splenic contraction [97–99]. Recently, a study established that T cells become sequestered in the bone marrow of both GBM patients and mice bearing glioma, in what has been proposed as a novel tumor immune-evasion tactic [98]. Such findings demonstrate an intriguing but troubling capacity of the tumor to modulate the systemic immune response despite its typical confinement within the brain.

Systemic immunosuppression is compounded by standard-of-care treatment in GBM patients. Following initial diagnosis, patients undergo maximal resection followed by radiotherapy and systemic temozolomide chemotherapy [1]. This treatment regimen damages the bone marrow and leads to severe myelosuppression in many patients [100]. The effects on immunity are compounded by the frequent addition of high-dose dexamethasone [101]. One study suggests that local chemotherapy may be a means of avoiding myelosuppression, as it increased lymphocyte number and potentiated the effects of anti-PD-1 [102]. Likewise, TMZ treatment may increase TMB in GBM patients, which may or may not be beneficial for ICB [103]. These studies do, however, illustrate the importance of considering standard-of-care treatments when administering immunotherapies.

New approaches/future directions

In view of ICB clinical trial failures for GBM, focus has now shifted toward understanding and addressing impediments to successful treatment [104]. A retrospective study examined 66 recurrent GBM patient samples, 17 of which were classified as responders. Using DNA, RNA, tissue imaging, and clinical data across several timepoints, Zhao et al. uncovered immune and genomic correlates of response to anti-PD-1 ICB in GBM. Of note, among nonresponders, *PTEN* was significantly more frequently mutated [105], which has been associated with resistance to lysis by tumor-specific T cells [106]. On the other hand, responders had molecular enrichment of MAPK pathway mutations, specifically PTPN11 and BRAF. Interestingly, BRAF mutations are common in melanoma patients that respond to immunotherapy and have been implicated in T-cell recognition of melanoma cells [107]. Analysis of immune infiltrates using quantitative multiplex immunofluorescence showed fewer $CD68^{+}$HLA-DR- macrophages and higher posttreatment CD8 T cells in responders when compared with nonresponders [105].

Two recent early-phase clinical trials administered anti-PD-1 therapy in the neoadjuvant setting (treatment before and after tumor resection) in patients with recurrent GBM [108, 109]. Schalper et al. conducted a single-arm phase II trial in 30 patients who received both presurgical and postsurgical nivolumab, until disease progression or unacceptable toxicity. No clinical benefit was observed in this trial although invaluable tumor biopsies before and after nivolumab treatment revealed the effects of treatment on the tumor microenvironment. These biopsies showed increased immune cell infiltration with more TCR clonal diversity when compared with historical controls, in addition to safety and feasibility of neoadjuvant ICB administration [109].

Similarly, Cloughesy et al. enrolled 35 recurrent GBM patients to a randomized open-label pilot study of pembrolizumab (anti-PD-1) to evaluate changes to the tumor microenvironment and survival following neoadjuvant and/or adjuvant therapy. Excitingly, overall survival in patients receiving neoadjuvant nivolumab with continued adjuvant administration had significantly improved survival compared with the adjuvant alone group. Further analysis using TCR sequencing, gene expression profiling, immunofluorescence, and mass cytometry revealed an upregulation of T-cell and IFN-γ-related gene signatures, and decreased enrichment of cell cycle/cancer proliferation signatures in the neoadjuvant cohort when compared with adjuvant alone [108]. These findings suggest that neoadjuvant ICB positively alters the tumor microenvironment with relevant immune cell activation and subsequent repression of cell cycle activity in tumor cells. They also suggest that there may yet be a promising path forward for ICB in GBM.

Conclusion

GBM remains universally lethal and one of the greatest therapeutic challenges in the immunotherapeutic era. GBM has armed itself impressively with means to deter an effective antitumor immune response. Mounting such a response begins with the activation of T cells and requires intact antigen presentation systems and the capacity to access tumor and maintain T-cell activity. Cancers, especially GBM, have evolved mechanisms to counter this process at nearly every step. Compounding few immunogenic GBM mutations with unrivaled tumor heterogeneity, T cells, already limited in number and function, stand little chance of meaningful activation. Even if arriving at tumor in a successfully activated state, T cells encounter T_{regs} and TAMs that restrict their function through the induction of tolerance. Should they sidestep tolerance, they are left with a tumor that still is able to provoke in them a severely exhausted phenotype, marked by the upregulation of inhibitory checkpoints. ICB has emerged as a therapeutic modality that might counter T-cell exhaustion and perpetuate T-cell activity, but the other challenges faced by T cells must still be addressed. Not surprisingly then, although ICB has been successful in many malignancies, success in GBM has not been realized thus far. Despite these initial failures, our improved understanding of the challenges facing T cells, as well as some favorable results in recent clinical trials, provides hope for successfully addressing the obstacles that may promote treatment resistance.

Acknowledgment

We would like to thank M.R. Llewellyn for the preparation of medical illustrations.

References

[1] Stupp R, Mason WP, van den Bent MJ, Weller M, Fisher B, Taphoorn MJ, Belanger K, Brandes AA, Marosi C, Bogdahn U, Curschmann J, Janzer RC, Ludwin SK, Gorlia T, Allgeier A, Lacombe D, Cairncross JG, Eisenhauer E, Mirimanoff RO, European Organisation for Research and Treatment of Cancer Brain Tumor and Radiotherapy Groups, National Cancer Institute of Canada Clinical Trials Group. Radiotherapy plus concomitant and adjuvant temozolomide for glioblastoma. N Engl J Med 2005;352(10):987–96.

[2] Stupp R, Taillibert S, Kanner A, Read W, Steinberg D, Lhermitte B, Toms S, Idbaih A, Ahluwalia MS, Fink K, Di Meco F, Lieberman F, Zhu JJ, Stragliotto G, Tran D, Brem S, Hottinger A, Kirson ED, Lavy-Shahaf G, Weinberg U, Kim CY, Paek SH, Nicholas G, Bruna J, Hirte H, Weller M, Palti Y, Hegi ME, Ram Z. Effect of tumor-treating fields plus maintenance temozolomide vs maintenance temozolomide alone on survival in patients with glioblastoma: a randomized clinical trial. JAMA 2017;318(23):2306–16.

[3] Gilbert MR, Dignam JJ, Armstrong TS, Wefel JS, Blumenthal DT, Vogelbaum MA, Colman H, Chakravarti A, Pugh S, Won M, Jeraj R, Brown PD, Jaeckle KA, Schiff D, Stieber VW, Brachman DG, Werner-Wasik M, Tremont-Lukats IW, Sulman EP, Aldape KD, Curran Jr WJ, Mehta MP. A randomized trial of bevacizumab for newly diagnosed glioblastoma. N Engl J Med 2014;370(8):699–708.

[4] Hodi FS, O'Day SJ, McDermott DF, Weber RW, Sosman JA, Haanen JB, Gonzalez R, Robert C, Schadendorf D, Hassel JC, Akerley W, van den Eertwegh AJ, Lutzky J, Lorigan P, Vaubel JM, Linette GP, Hogg D, Ottensmeier CH, Lebbe C, Peschel C, Quirt I, Clark JI, Wolchok JD, Weber JS, Tian J, Yellin MJ, Nichol GM, Hoos A, Urba WJ. Improved survival with ipilimumab in patients with metastatic melanoma. N Engl J Med 2010;363(8):711–23.

[5] Filley AC, Henriquez M, Dey M. Recurrent glioma clinical trial, CheckMate-143: the game is not over yet. Oncotarget 2017;8(53):91779–94.

[6] Hellmann MD, Ciuleanu TE, Pluzanski A, Lee JS, Otterson GA, Audigier-Valette C, Minenza E, Linardou H, Burgers S, Salman P, Borghaei H, Ramalingam SS, Brahmer J, Reck M, O'Byrne KJ, Geese WJ, Green G, Chang

H, Szustakowski J, Bhagavatheeswaran P, Healey D, Fu Y, Nathan F, Paz-Ares L. Nivolumab plus ipilimumab in lung cancer with a high tumor mutational burden. N Engl J Med 2018;378(22):2093–104.
[7] Lemery S, Keegan P, Pazdur R. First FDA approval agnostic of cancer site—when a biomarker defines the indication. N Engl J Med 2017;377(15):1409–12.
[8] Lipson EJ, Drake CG. Ipilimumab: an anti-CTLA-4 antibody for metastatic melanoma. Clin Cancer Res 2011;17 (22):6958–62.
[9] Reardon DA, Omuro A, Brandes AA, Rieger J, Wick A, Sepulveda J, Phuphanich S, de Souza P, Ahluwalia MS, Lim M, Vlahovic G, Sampson J. OS10.3 randomized phase 3 study evaluating the efficacy and safety of nivolumab vs bevacizumab in patients with recurrent glioblastoma: CheckMate 143. Neuro-Oncology 2017;19(suppl_3):iii21.
[10] Goral S. The three-signal hypothesis of lymphocyte activation/targets for immunosuppression. Dial Transplant 2011;40(1):14–6.
[11] Zhu Y, Yao S, Chen L. Cell surface signaling molecules in the control of immune responses: a tide model. Immunity 2011;34(4):466–78.
[12] Dunn GP, Old LJ, Schreiber RD. The three Es of cancer immunoediting. Annu Rev Immunol 2004;22:329–60.
[13] Vasilevko V, Ghochikyan A, Holterman MJ, Agadjanyan MG. CD80 (B7-1) and CD86 (B7-2) are functionally equivalent in the initiation and maintenance of CD4+ T-cell proliferation after activation with suboptimal doses of PHA. DNA Cell Biol 2002;21(3):137–49.
[14] Chen L, Flies DB. Molecular mechanisms of T cell co-stimulation and co-inhibition. Nat Rev Immunol 2013;13 (4):227–42.
[15] Linsley PS, Bradshaw J, Greene J, Peach R, Bennett KL, Mittler RS. Intracellular trafficking of CTLA-4 and focal localization towards sites of TCR engagement. Immunity 1996;4(6):535–43.
[16] Qian J, Wang C, Wang B, Yang J, Wang Y, Luo F, Xu J, Zhao C, Liu R, Chu Y. The IFN-gamma/PD-L1 axis between T cells and tumor microenvironment: hints for glioma anti-PD-1/PD-L1 therapy. J Neuroinflammation 2018;15(1):290.
[17] Wang ZL, Li GZ, Wang QW, Bao ZS, Wang Z, Zhang CB, Jiang T. PD-L2 expression is correlated with the molecular and clinical features of glioma, and acts as an unfavorable prognostic factor. Onco Targets Ther 2019;8 (2), e1541535.
[18] Phillips EA, Tocheva AS, Mor A, Kong X. Structural, functional, and evolutionary differences between PD-L1 and PD-L2. J Immunol 2018;200(1).
[19] Driessens G, Kline J, Gajewski TF. Costimulatory and coinhibitory receptors in anti-tumor immunity. Immunol Rev 2009;229(1):126–44.
[20] Lucca LE, Hafler DA. Co-inhibitory blockade while preserving tolerance: checkpoint inhibitors for glioblastoma. Immunol Rev 2017;276(1):9–25.
[21] Lozano E, Dominguez-Villar M, Kuchroo V, Hafler DA. The TIGIT/CD226 axis regulates human T cell function. J Immunol 2012;188(8):3869–75.
[22] Woroniecka KI, Rhodin KE, Chongsathidkiet P, Keith KA, Fecci PE. T-cell dysfunction in glioblastoma: applying a new framework. Clin Cancer Res 2018;24(16):3792–802.
[23] Koyama S, Akbay EA, Li YY, Herter-Sprie GS, Buczkowski KA, Richards WG, Gandhi L, Redig AJ, Rodig SJ, Asahina H, Jones RE, Kulkarni MM, Kuraguchi M, Palakurthi S, Fecci PE, Johnson BE, Janne PA, Engelman JA, Gangadharan SP, Costa DB, Freeman GJ, Bueno R, Hodi FS, Dranoff G, Wong KK, Hammerman PS. Adaptive resistance to therapeutic PD-1 blockade is associated with upregulation of alternative immune checkpoints. Nat Commun 2016;7:10501.
[24] Woroniecka K, Fecci PE. T-cell exhaustion in glioblastoma. Oncotarget 2018;9(82):35287–8.
[25] Leach DR, Krummel MF, Allison JP. Enhancement of antitumor immunity by CTLA-4 blockade. Science 1996;271(5256):1734–6.
[26] Kwon ED, Hurwitz AA, Foster BA, Madias C, Feldhaus AL, Greenberg NM, Burg MB, Allison JP. Manipulation of T cell costimulatory and inhibitory signals for immunotherapy of prostate cancer. Proc Natl Acad Sci U S A 1997;94(15):8099–103.
[27] Jure-Kunkel M, Masters G, Girit E, Dito G, Lee F, Hunt JT, Humphrey R. Synergy between chemotherapeutic agents and CTLA-4 blockade in preclinical tumor models. Cancer Immunol Immunother 2013;62(9):1533–45.
[28] Hurwitz AA, Yu TF, Leach DR, Allison JP. CTLA-4 blockade synergizes with tumor-derived granulocyte-macrophage colony-stimulating factor for treatment of an experimental mammary carcinoma. Proc Natl Acad Sci U S A 1998;95(17):10067–71.

[29] Shrikant P, Khoruts A, Mescher MF. CTLA-4 blockade reverses CD8+ T cell tolerance to tumor by a CD4+ T cell- and IL-2-dependent mechanism. Immunity 1999;11(4):483–93.

[30] van Elsas A, Hurwitz AA, Allison JP. Combination immunotherapy of B16 melanoma using anti-cytotoxic T lymphocyte-associated antigen 4 (CTLA-4) and granulocyte/macrophage colony-stimulating factor (GM-CSF)-producing vaccines induces rejection of subcutaneous and metastatic tumors accompanied by autoimmune depigmentation. J Exp Med 1999;190(3):355–66.

[31] Iwai Y, Terawaki S, Honjo T. PD-1 blockade inhibits hematogenous spread of poorly immunogenic tumor cells by enhanced recruitment of effector T cells. Int Immunol 2005;17(2):133–44.

[32] Nomi T, Sho M, Akahori T, Hamada K, Kubo A, Kanehiro H, Nakamura S, Enomoto K, Yagita H, Azuma M, Nakajima Y. Clinical significance and therapeutic potential of the programmed death-1 ligand/programmed death-1 pathway in human pancreatic cancer. Clin Cancer Res 2007;13(7):2151–7.

[33] Hirano F, Kaneko K, Tamura H, Dong H, Wang S, Ichikawa M, Rietz C, Flies DB, Lau JS, Zhu G, Tamada K, Chen L. Blockade of B7-H1 and PD-1 by monoclonal antibodies potentiates cancer therapeutic immunity. Cancer Res 2005;65(3):1089–96.

[34] Strome SE, Dong H, Tamura H, Voss SG, Flies DB, Tamada K, Salomao D, Cheville J, Hirano F, Lin W, Kasperbauer JL, Ballman KV, Chen L. B7-H1 blockade augments adoptive T-cell immunotherapy for squamous cell carcinoma. Cancer Res 2003;63(19):6501–5.

[35] Hodi FS, Mihm MC, Soiffer RJ, Haluska FG, Butler M, Seiden MV, Davis T, Henry-Spires R, MacRae S, Willman A, Padera R, Jaklitsch MT, Shankar S, Chen TC, Korman A, Allison JP, Dranoff G. Biologic activity of cytotoxic T lymphocyte-associated antigen 4 antibody blockade in previously vaccinated metastatic melanoma and ovarian carcinoma patients. Proc Natl Acad Sci U S A 2003;100(8):4712–7.

[36] Brahmer JR, Drake CG, Wollner I, Powderly JD, Picus J, Sharfman WH, Stankevich E, Pons A, Salay TM, McMiller TL, Gilson MM, Wang C, Selby M, Taube JM, Anders R, Chen L, Korman AJ, Pardoll DM, Lowy I, Topalian SL. Phase I study of single-agent anti-programmed death-1 (MDX-1106) in refractory solid tumors: safety, clinical activity, pharmacodynamics, and immunologic correlates. J Clin Oncol 2010;28(19):3167–75.

[37] Topalian SL, Hodi FS, Brahmer JR, Gettinger SN, Smith DC, McDermott DF, Powderly JD, Carvajal RD, Sosman JA, Atkins MB, Leming PD, Spigel DR, Antonia SJ, Horn L, Drake CG, Pardoll DM, Chen L, Sharfman WH, Anders RA, Taube JM, McMiller TL, Xu H, Korman AJ, Jure-Kunkel M, Agrawal S, McDonald D, Kollia GD, Gupta A, Wigginton JM, Sznol M. Safety, activity, and immune correlates of anti-PD-1 antibody in cancer. N Engl J Med 2012;366(26):2443–54.

[38] Woroniecka K, Chongsathidkiet P, Rhodin K, Kemeny H, Dechant C, Farber SH, Elsamadicy AA, Cui X, Koyama S, Jackson C, Hansen LJ, Johanns TM, Sanchez-Perez L, Chandramohan V, Yu YA, Bigner DD, Giles A, Healy P, Dranoff G, Weinhold KJ, Dunn GP, Fecci PE. T-cell exhaustion signatures vary with tumor type and are severe in glioblastoma. Clin Cancer Res 2018;24(17):4175–86.

[39] Fecci PE, Ochiai H, Mitchell DA, Grossi PM, Sweeney AE, Archer GE, Cummings T, Allison JP, Bigner DD, Sampson JH. Systemic CTLA-4 blockade ameliorates glioma-induced changes to the CD4+ T cell compartment without affecting regulatory T-cell function. Clin Cancer Res 2007;13(7):2158–67.

[40] Reardon DA, Gokhale PC, Klein SR, Ligon KL, Rodig SJ, Ramkissoon SH, Jones KL, Conway AS, Liao X, Zhou J, Wen PY, Van Den Abbeele AD, Hodi FS, Qin L, Kohl NE, Sharpe AH, Dranoff G, Freeman GJ. Glioblastoma eradication following Immune checkpoint blockade in an orthotopic, immunocompetent model. Cancer Immunol Res 2016;4(2):124–35.

[41] Wainwright DA, Chang AL, Dey M, Balyasnikova IV, Kim CK, Tobias A, Cheng Y, Kim JW, Qiao J, Zhang L, Han Y, Lesniak MS. Durable therapeutic efficacy utilizing combinatorial blockade against IDO, CTLA-4, and PD-L1 in mice with brain tumors. Clin Cancer Res 2014;20(20):5290–301.

[42] Zeng J, See AP, Phallen J, Jackson CM, Belcaid Z, Ruzevick J, Durham N, Meyer C, Harris TJ, Albesiano E, Pradilla G, Ford E, Wong J, Hammers HJ, Mathios D, Tyler B, Brem H, Tran PT, Pardoll D, Drake CG, Lim M. -Anti-PD-1 blockade and stereotactic radiation produce long-term survival in mice with intracranial gliomas. Int J Radiat Oncol Biol Phys 2013;86(2):343–9.

[43] Kosaka A, Ohkuri T, Okada H. Combination of an agonistic anti-CD40 monoclonal antibody and the COX-2 inhibitor celecoxib induces anti-glioma effects by promotion of type-1 immunity in myeloid cells and T-cells. Cancer Immunol Immunother 2014;63(8):847–57.

[44] Guan X, Hasan MN, Begum G, Kohanbash G, Carney KE, Pigott VM, Persson AI, Castro MG, Jia W, Sun D. Blockade of Na/H exchanger stimulates glioma tumor immunogenicity and enhances combinatorial TMZ and anti-PD-1 therapy. Cell Death Dis 2018;9(10):1010.

[45] Reardon DA, Sampson JH, Sahebjam S, Lim M, Baehring JM, Vlahovic G, Cloughesy TF, Strauss LC, Latek RR, Paliwal P, Harbison CT, Voloschin AD, Omuro AMP. Safety and activity of nivolumab (nivo) monotherapy and nivo in combination with ipilimumab (ipi) in recurrent glioblastoma (GBM): updated results from checkmate-143. J Clin Oncol 2016;34(15_suppl):2014.

[46] Chamberlain MC, Kim BT. Nivolumab for patients with recurrent glioblastoma progressing on bevacizumab: a retrospective case series. J Neuro-Oncol 2017;133(3):561–9.

[47] Reardon DA, Kim T-M, Frenel J-S, Santoro A, Lopez J, Subramaniam DS, Siu LL, Rodon J, Tamura K, Saraf S, Morosky A, Stein K, Soria J-C. ATIM-35. results of the phase IB keynote-028 multi-cohort trial of pembrolizumab monotherapy in patients with recurrent PD-L1-positive glioblastoma multiforme (GBM). Neuro-Oncology 2016;18(suppl_6):vi25–6.

[48] Reiss SN, Yerram P, Modelevsky L, Grommes C. Retrospective review of safety and efficacy of programmed cell death-1 inhibitors in refractory high grade gliomas. J Immunother Cancer 2017;5(1):99.

[49] Reardon DA, Nayak L, Peters KB, Clarke JL, Jordan JT, Groot JFD, Nghiemphu PL, Kaley TJ, Colman H, Gaffey SC, Caruso V, Debruyne MB, Bhavsar C, Molinaro AM, Smith T, Severgnini M, Wen PY. Phase II study of pembrolizumab or pembrolizumab plus bevacizumab for recurrent glioblastoma (rGBM) patients. J Clin Oncol 2018;36(15_suppl):2006.

[50] Groot JFD, Penas-Prado M, Mandel JJ, O'Brien BJ, Weathers S-PS, Zhou S, Hunter K, Alfaro-Munoz K, Fuller GN, Huse J, Rao G, Weinberg JS, Prabhu SS, Ferguson SD, Yuan Y, Vence LM, Allison JP, Sharma P, Heimberger AB. Window-of-opportunity clinical trial of a PD-1 inhibitor in patients with recurrent glioblastoma. J Clin Oncol 2018;36(15_suppl):2008.

[51] Reardon DA, Kaley TJ, Dietrich J, Clarke JL, Dunn GP, Lim M, Cloughesy TF, Gan HK, Park AJ, Schwarzenberger P, Ricciardi T, Macri MJ, Ryan A, Venhaus RR. Phase 2 study to evaluate safety and efficacy of MEDI4736 (durvalumab [DUR]) in glioblastoma (GBM) patients: an update. J Clin Oncol 2017;35(15_suppl):2042.

[52] Reardon DA, Kaley TJ, Dietrich J, Clarke JL, Dunn G, Lim M, Cloughesy TF, Gan HK, Park AJ, Schwarzenberger P, Ricciardi T, Macri MJ, Ryan A, Venhaus RR. Phase II study to evaluate safety and efficacy of MEDI4736 (durvalumab) + radiotherapy in patients with newly diagnosed unmethylated MGMT glioblastoma (new unmeth GBM). J Clin Oncol 2019;37(15_suppl):2032.

[53] Coulie PG, Van den Eynde BJ, van der Bruggen P, Boon T. Tumour antigens recognized by T lymphocytes: at the core of cancer immunotherapy. Nat Rev Cancer 2014;14(2):135–46.

[54] Havel JJ, Chowell D, Chan TA. The evolving landscape of biomarkers for checkpoint inhibitor immunotherapy. Nat Rev Cancer 2019;19(3):133–50.

[55] Gubin MM, Zhang X, Schuster H, Caron E, Ward JP, Noguchi T, Ivanova Y, Hundal J, Arthur CD, Krebber WJ, Mulder GE, Toebes M, Vesely MD, Lam SS, Korman AJ, Allison JP, Freeman GJ, Sharpe AH, Pearce EL, Schumacher TN, Aebersold R, Rammensee HG, Melief CJ, Mardis ER, Gillanders WE, Artyomov MN, Schreiber RD. Checkpoint blockade cancer immunotherapy targets tumour-specific mutant antigens. Nature 2014;515 (7528):577–81.

[56] Lennerz V, Fatho M, Gentilini C, Frye RA, Lifke A, Ferel D, Wolfel C, Huber C, Wolfel T. The response of autologous T cells to a human melanoma is dominated by mutated neoantigens. Proc Natl Acad Sci U S A 2005;102(44):16013–8.

[57] Dubey P, Hendrickson RC, Meredith SC, Siegel CT, Shabanowitz J, Skipper JC, Engelhard VH, Hunt DF, Schreiber H. The immunodominant antigen of an ultraviolet-induced regressor tumor is generated by a somatic point mutation in the DEAD box helicase p68. J Exp Med 1997;185(4):695–705.

[58] Snyder A, Makarov V, Merghoub T, Yuan J, Zaretsky JM, Desrichard A, Walsh LA, Postow MA, Wong P, Ho TS, Hollmann TJ, Bruggeman C, Kannan K, Li Y, Elipenahli C, Liu C, Harbison CT, Wang L, Ribas A, Wolchok JD, Chan TA. Genetic basis for clinical response to CTLA-4 blockade in melanoma. N Engl J Med 2014;371(23): 2189–99.

[59] Rizvi NA, Hellmann MD, Snyder A, Kvistborg P, Makarov V, Havel JJ, Lee W, Yuan J, Wong P, Ho TS, Miller ML, Rekhtman N, Moreira AL, Ibrahim F, Bruggeman C, Gasmi B, Zappasodi R, Maeda Y, Sander C, Garon EB, Merghoub T, Wolchok JD, Schumacher TN, Chan TA. Cancer immunology. Mutational landscape determines sensitivity to PD-1 blockade in non-small cell lung cancer. Science 2015;348(6230):124–8.

[60] Alexandrov LB, Nik-Zainal S, Wedge DC, Aparicio SA, Behjati S, Biankin AV, Bignell GR, Bolli N, Borg A, Borresen-Dale AL, Boyault S, Burkhardt B, Butler AP, Caldas C, Davies HR, Desmedt C, Eils R, Eyfjord JE, Foekens JA, Greaves M, Hosoda F, Hutter B, Ilicic T, Imbeaud S, Imielinski M, Jager N, Jones DT, Jones D,

Knappskog S, Kool M, Lakhani SR, Lopez-Otin C, Martin S, Munshi NC, Nakamura H, Northcott PA, Pajic M, Papaemmanuil E, Paradiso A, Pearson JV, Puente XS, Raine K, Ramakrishna M, Richardson AL, Richter J, Rosenstiel P, Schlesner M, Schumacher TN, Span PN, Teague JW, Totoki Y, Tutt AN, Valdes-Mas R, van Buuren MM, van 't Veer L, Vincent-Salomon A, Waddell N, Yates LR, Australian Pancreatic Cancer Genome Initiative, ICGC Breast Cancer Consortium, ICGC MMML-Seq Consortium, ICGC PedBrain, Zucman-Rossi J, Futreal PA, McDermott U, Lichter P, Meyerson M, Grimmond SM, Siebert R, Campo E, Shibata T, Pfister SM, Campbell PJ, Stratton MR. Signatures of mutational processes in human cancer. Nature 2013;500(7463):415–21.

[61] Schumacher TN, Schreiber RD. Neoantigens in cancer immunotherapy. Science 2015;348(6230):69–74.

[62] Champiat S, Ferte C, Lebel-Binay S, Eggermont A, Soria JC. Exomics and immunogenics: bridging mutational load and immune checkpoints efficacy. Onco Targets Ther 2014;3(1), e27817.

[63] Hellmann MD, Callahan MK, Awad MM, Calvo E, Ascierto PA, Atmaca A, Rizvi NA, Hirsch FR, Selvaggi G, Szustakowski JD, Sasson A, Golhar R, Vitazka P, Chang H, Geese WJ, Antonia SJ. Tumor mutational burden and efficacy of nivolumab monotherapy and in combination with ipilimumab in small-cell lung cancer. Cancer Cell 2018;33(5):853–61. e4.

[64] Colli LM, Machiela MJ, Myers TA, Jessop L, Yu K, Chanock SJ. Burden of nonsynonymous mutations among TCGA cancers and candidate immune checkpoint inhibitor responses. Cancer Res 2016;76(13):3767–72.

[65] Hodges TR, Ott M, Xiu J, Gatalica Z, Swensen J, Zhou S, Huse JT, de Groot J, Li S, Overwijk WW, Spetzler D, Heimberger AB. Mutational burden, immune checkpoint expression, and mismatch repair in glioma: implications for immune checkpoint immunotherapy. Neuro-Oncology 2017;19(8):1047–57.

[66] Bouffet E, Larouche V, Campbell BB, Merico D, de Borja R, Aronson M, Durno C, Krueger J, Cabric V, Ramaswamy V, Zhukova N, Mason G, Farah R, Afzal S, Yalon M, Rechavi G, Magimairajan V, Walsh MF, Constantini S, Dvir R, Elhasid R, Reddy A, Osborn M, Sullivan M, Hansford J, Dodgshun A, Klauber-Demore N, Peterson L, Patel S, Lindhorst S, Atkinson J, Cohen Z, Laframboise R, Dirks P, Taylor M, Malkin D, Albrecht S, Dudley RW, Jabado N, Hawkins CE, Shlien A, Tabori U. Immune checkpoint inhibition for hypermutant glioblastoma multiforme resulting from germline biallelic mismatch repair deficiency. J Clin Oncol 2016;34 (19):2206–11.

[67] Lee V, Murphy A, Le DT, Diaz Jr LA. Mismatch repair deficiency and response to immune checkpoint blockade. Oncologist 2016;21(10):1200–11.

[68] Qazi MA, Vora P, Venugopal C, Sidhu SS, Moffat J, Swanton C, Singh SK. Intratumoral heterogeneity: pathways to treatment resistance and relapse in human glioblastoma. Ann Oncol 2017;28(7):1448–56.

[69] Patel AP, Tirosh I, Trombetta JJ, Shalek AK, Gillespie SM, Wakimoto H, Cahill DP, Nahed BV, Curry WT, Martuza RL, Louis DN, Rozenblatt-Rosen O, Suva ML, Regev A, Bernstein BE. Single-cell RNA-seq highlights intratumoral heterogeneity in primary glioblastoma. Science 2014;344(6190):1396–401.

[70] Johnson BE, Mazor T, Hong C, Barnes M, Aihara K, McLean CY, Fouse SD, Yamamoto S, Ueda H, Tatsuno K, Asthana S, Jalbert LE, Nelson SJ, Bollen AW, Gustafson WC, Charron E, Weiss WA, Smirnov IV, Song JS, Olshen AB, Cha S, Zhao Y, Moore RA, Mungall AJ, Jones SJM, Hirst M, Marra MA, Saito N, Aburatani H, Mukasa A, Berger MS, Chang SM, Taylor BS, Costello JF. Mutational analysis reveals the origin and therapy-driven evolution of recurrent glioma. Science 2014;343(6167):189–93.

[71] Reinartz R, Wang S, Kebir S, Silver DJ, Wieland A, Zheng T, Kupper M, Rauschenbach L, Fimmers R, Shepherd TM, Trageser D, Till A, Schafer N, Glas M, Hillmer AM, Cichon S, Smith AA, Pietsch T, Liu Y, Reynolds BA, Yachnis A, Pincus DW, Simon M, Brustle O, Steindler DA, Scheffler B. Functional subclone profiling for prediction of treatment-induced intratumor population shifts and discovery of rational drug combinations in human glioblastoma. Clin Cancer Res 2017;23(2):562–74.

[72] Galluzzi L, Chan TA, Kroemer G, Wolchok JD, Lopez-Soto A. The hallmarks of successful anticancer immunotherapy. Sci Transl Med 2018;10(459).

[73] Fridman WH, Zitvogel L, Sautes-Fridman C, Kroemer G. The immune contexture in cancer prognosis and treatment. Nat Rev Clin Oncol 2017;14(12):717–34.

[74] Salmon H, Remark R, Gnjatic S, Merad M. Host tissue determinants of tumour immunity. Nat Rev Cancer 2019;19(4):215–27.

[75] Robert L, Harview C, Emerson R, Wang X, Mok S, Homet B, Comin-Anduix B, Koya RC, Robins H, Tumeh PC, Ribas A. Distinct immunological mechanisms of CTLA-4 and PD-1 blockade revealed by analyzing TCR usage in blood lymphocytes. Onco Targets Ther 2014;3, e29244.

[76] Heppt MV, Steeb T, Schlager JG, Rosumeck S, Dressler C, Ruzicka T, Nast A, Berking C. Immune checkpoint blockade for unresectable or metastatic uveal melanoma: a systematic review. Cancer Treat Rev 2017;60:44–52.

[77] Huang AC, Postow MA, Orlowski RJ, Mick R, Bengsch B, Manne S, Xu W, Harmon S, Giles JR, Wenz B, Adamow M, Kuk D, Panageas KS, Carrera C, Wong P, Quagliarello F, Wubbenhorst B, D'Andrea K, Pauken KE, Herati RS, Staupe RP, Schenkel JM, McGettigan S, Kothari S, George SM, Vonderheide RH, Amaravadi RK, Karakousis GC, Schuchter LM, Xu X, Nathanson KL, Wolchok JD, Gangadhar TC, Wherry EJ. T-cell invigoration to tumour burden ratio associated with anti-PD-1 response. Nature 2017;545(7652):60–5.
[78] Fecci PE, Mitchell DA, Whitesides JF, Xie W, Friedman AH, Archer GE, Herndon 2nd JE, Bigner DD, Dranoff G, Sampson JH. Increased regulatory T-cell fraction amidst a diminished CD4 compartment explains cellular immune defects in patients with malignant glioma. Cancer Res 2006;66(6):3294–302.
[79] Tomaszewski W, Sanchez-Perez L, Gajewski TF, Sampson JH. Brain tumor microenvironment and host state: implications for immunotherapy. Clin Cancer Res 2019;25(14):4202–10.
[80] Thorsson V, Gibbs DL, Brown SD, Wolf D, Bortone DS, Yang THO, Porta-Pardo E, Gao GF, Plaisier CL, Eddy JA, Ziv E, Culhane AC, Paull EO, Sivakumar IKA, Gentles AJ, Malhotra R, Farshidfar F, Colaprico A, Parker JS, Mose LE, Vo NS, Liu J, Liu Y, Rader J, Dhankani V, Reynolds SM, Bowlby R, Califano A, Cherniack AD, Anastassiou D, Bedognetti D, Rao A, Chen K, Krasnitz A, Hu H, Malta TM, Noushmehr H, Pedamallu CS, Bullman S, Ojesina AI, Lamb A, Zhou W, Shen H, Choueiri TK, Weinstein JN, Guinney J, Saltz J, Holt RA, Rabkin CE, N. Cancer Genome Atlas Research, Lazar AJ, Serody JS, Demicco EG, Disis ML, Vincent BG, Shmulevich L. The immune landscape of cancer. Immunity 2018;48(4):812–30. e14.
[81] Ausiello CM, Palma C, Maleci A, Spagnoli GC, Amici C, Antonelli G, Casciani CU, Cassone A. Cell mediated cytotoxicity and cytokine production in peripheral blood mononuclear cells of glioma patients. Eur J Cancer 1991;27(5):646–50.
[82] Elliott LH, Brooks WH, Roszman TL. Cytokinetic basis for the impaired activation of lymphocytes from patients with primary intracranial tumors. J Immunol 1984;132(3):1208–15.
[83] El Andaloussi A, Han Y, Lesniak MS. Prolongation of survival following depletion of CD4+CD25+ regulatory T cells in mice with experimental brain tumors. J Neurosurg 2006;105(3):430–7.
[84] El Andaloussi A, Sonabend AM, Han Y, Lesniak MS. Stimulation of TLR9 with CpG ODN enhances apoptosis of glioma and prolongs the survival of mice with experimental brain tumors. Glia 2006;54(6):526–35.
[85] Learn CA, Fecci PE, Schmittling RJ, Xie W, Karikari I, Mitchell DA, Archer GE, Wei Z, Dressman H, Sampson JH. Profiling of CD4+, CD8+, and CD4+CD25+CD45RO+FoxP3+ T cells in patients with malignant glioma reveals differential expression of the immunologic transcriptome compared with T cells from healthy volunteers. Clin Cancer Res 2006;12(24):7306–15.
[86] Chen DS, Mellman I. Oncology meets immunology: the cancer-immunity cycle. Immunity 2013;39(1):1–10.
[87] Pardoll DM. The blockade of immune checkpoints in cancer immunotherapy. Nat Rev Cancer 2012;12(4): 252–64.
[88] Wherry EJ, Kurachi M. Molecular and cellular insights into T cell exhaustion. Nat Rev Immunol 2015;15(8): 486–99.
[89] Medawar PB. Immunity to homologous grafted skin; the fate of skin homografts transplanted to the brain, to subcutaneous tissue, and to the anterior chamber of the eye. Br J Exp Pathol 1948;29(1):58–69.
[90] Galea I, Bechmann I, Perry VH. What is immune privilege (not)? Trends Immunol 2007;28(1):12–8.
[91] Cserr HF, Harling-Berg CJ, Knopf PM. Drainage of brain extracellular fluid into blood and deep cervical lymph and its immunological significance. Brain Pathol 1992;2(4):269–76.
[92] Louveau A, Smirnov I, Keyes TJ, Eccles JD, Rouhani SJ, Peske JD, Derecki NC, Castle D, Mandell JW, Lee KS, Harris TH, Kipnis J. Structural and functional features of central nervous system lymphatic vessels. Nature 2015;523(7560):337–41.
[93] Schlager C, Korner H, Krueger M, Vidoli S, Haberl M, Mielke D, Brylla E, Issekutz T, Cabanas C, Nelson PJ, Ziemssen T, Rohde V, Bechmann I, Lodygin D, Odoardi F, Flugel A. Effector T-cell trafficking between the leptomeninges and the cerebrospinal fluid. Nature 2016;530(7590):349–53.
[94] Cserr HF, Knopf PM. Cervical lymphatics, the blood-brain barrier and the immunoreactivity of the brain: a new view. Immunol Today 1992;13(12):507–12.
[95] Long DM. Capillary ultrastructure and the blood-brain barrier in human malignant brain tumors. J Neurosurg 1970;32(2):127–44.
[96] Watkins S, Robel S, Kimbrough IF, Robert SM, Ellis-Davies G, Sontheimer H. Disruption of astrocyte-vascular coupling and the blood-brain barrier by invading glioma cells. Nat Commun 2014;5:4196.
[97] Brooks WH, Roszman TL, Mahaley MS, Woosley RE. Immunobiology of primary intracranial tumours. II. Analysis of lymphocyte subpopulations in patients with primary brain tumours. Clin Exp Immunol 1977;29(1):61–6.

[98] Chongsathidkiet P, Jackson C, Koyama S, Loebel F, Cui X, Farber SH, Woroniecka K, Elsamadicy AA, Dechant CA, Kemeny HR, Sanchez-Perez L, Cheema TA, Souders NC, Herndon JE, Coumans JV, Everitt JI, Nahed BV, Sampson JH, Gunn MD, Martuza RL, Dranoff G, Curry WT, Fecci PE. Sequestration of T cells in bone marrow in the setting of glioblastoma and other intracranial tumors. Nat Med 2018;24(9):1459–68.

[99] Dix AR, Brooks WH, Roszman TL, Morford LA. Immune defects observed in patients with primary malignant brain tumors. J Neuroimmunol 1999;100(1–2):216–32.

[100] Lombardi G, Rumiato E, Bertorelle R, Saggioro D, Farina P, Della Puppa A, Zustovich F, Berti F, Sacchetto V, Marcato R, Amadori A, Zagonel V. Clinical and genetic factors associated with severe hematological toxicity in glioblastoma patients during radiation plus temozolomide treatment: a prospective study. Am J Clin Oncol 2015;38(5):514–9.

[101] Giles AJ, Hutchinson MND, Sonnemann HM, Jung J, Fecci PE, Ratnam NM, Zhang W, Song H, Bailey R, Davis D, Reid CM, Park DM, Gilbert MR. Dexamethasone-induced immunosuppression: mechanisms and implications for immunotherapy. J Immunother Cancer 2018;6(1):51.

[102] Mathios D, Kim JE, Mangraviti A, Phallen J, Park CK, Jackson CM, Garzon-Muvdi T, Kim E, Theodros D, Polanczyk M, Martin AM, Suk I, Ye X, Tyler B, Bettegowda C, Brem H, Pardoll DM, Lim M. Anti-PD-1 antitumor immunity is enhanced by local and abrogated by systemic chemotherapy in GBM. Sci Transl Med 2016;8(370):370ra180.

[103] Daniel P, Sabri S, Chaddad A, Meehan B, Jean-Claude B, Rak J, Abdulkarim BS. Temozolomide induced hypermutation in glioma: evolutionary mechanisms and therapeutic opportunities. Front Oncol 2019;9:41.

[104] Ito H, Nakashima H, Chiocca EA. Molecular responses to immune checkpoint blockade in glioblastoma. Nat Med 2019;25(3):359–61.

[105] Zhao J, Chen AX, Gartrell RD, Silverman AM, Aparicio L, Chu T, Bordbar D, Shan D, Samanamud J, Mahajan A, Filip I, Orenbuch R, Goetz M, Yamaguchi JT, Cloney M, Horbinski C, Lukas RV, Raizer J, Rae AI, Yuan J, Canoll P, Bruce JN, Saenger YM, Sims P, Iwamoto FM, Sonabend AM, Rabadan R. Immune and genomic correlates of response to anti-PD-1 immunotherapy in glioblastoma. Nat Med 2019;25(3):462–9.

[106] Parsa AT, Waldron JS, Panner A, Crane CA, Parney IF, Barry JJ, Cachola KE, Murray JC, Tihan T, Jensen MC, Mischel PS, Stokoe D, Pieper RO. Loss of tumor suppressor PTEN function increases B7-H1 expression and immunoresistance in glioma. Nat Med 2007;13(1):84–8.

[107] Pan D, Kobayashi A, Jiang P, de Andrade LF, Tay RE, Luoma AM, Tsoucas D, Qiu X, Lim K, Rao P, Long HW, Yuan GC, Doench J, Brown M, Liu XS, Wucherpfennig KW. A major chromatin regulator determines resistance of tumor cells to T cell-mediated killing. Science 2018;359(6377):770–5.

[108] Cloughesy TF, Mochizuki AY, Orpilla JR, Hugo W, Lee AH, Davidson TB, Wang AC, Ellingson BM, Rytlewski JA, Sanders CM, Kawaguchi ES, Du L, Li G, Yong WH, Gaffey SC, Cohen AL, Mellinghoff IK, Lee EQ, Reardon DA, O'Brien BJ, Butowski NA, Nghiemphu PL, Clarke JL, Arrillaga-Romany IC, Colman H, Kaley TJ, de Groot JF, Liau LM, Wen PY, Prins RM. Neoadjuvant anti-PD-1 immunotherapy promotes a survival benefit with intratumoral and systemic immune responses in recurrent glioblastoma. Nat Med 2019;25(3):477–86.

[109] Schalper KA, Rodriguez-Ruiz ME, Diez-Valle R, Lopez-Janeiro A, Porciuncula A, Idoate MA, Inoges S, de Andrea C, de Cerio AL-D, Tejada S, Berraondo P, Villarroel-Espindola F, Choi J, Gurpide A, Giraldez M, Goicoechea I, Perez-Larraya JG, Sanmamed MF, Perez-Gracia JL, Melero I. Neoadjuvant nivolumab modifies the tumor immune microenvironment in resectable glioblastoma. Nat Med 2019;25(3):470–6.

CHAPTER

7

Immunomodulatory roles of myeloid cells in gliomas

Pavlina Chuntova[a,*], *Payal B. Watchmaker*[a,*], *Ryan Gilbert*[a,*], *and Hideho Okada*[a,b,c]

[a]Department of Neurological Surgery, University of California, San Francisco, CA, United States
[b]The Parker Institute for Cancer Immunotherapy, San Francisco, CA, United States
[c]Cancer Immunotherapy Program, University of California, San Francisco, CA, United States

Abstract

Myeloid cells represent the most abundant immune cell-type infiltrating glioblastoma (GBM). These myeloid cells are a heterogeneous population with primarily tumor-promoting functions. Macrophage activation status has been shown to span a broad spectrum of phenotypes that are highly dependent on the stimuli present in a specific tissue microenvironment. The paradigm defining macrophages as either classically activated antitumor M1 macrophages or immunosuppressive M2 macrophages provides points of reference; however, recent data indicate the differentiation of GBM-infiltrating macrophages is a continuous process, characterized by the large numbers of relatively immature myeloid cells that can express both canonical M1 and M2 markers at the same time. These cells are likely to contribute to the outcomes of GBM patients treated with immune therapies given their high abundance and immunosuppressive functions.

Abbreviations

AHR	aryl hydrocarbon receptor
ARG1	arginase I
CECR1	cat eye syndrome critical region protein 1
CNS	central nervous system
CSF-1	colony-stimulating factor 1
DCs	dendritic cells
DIPG	diffuse intrinsic pontine glioma
EGF	epidermal growth factor

[*] These authors contributed equally.

https://doi.org/10.1016/B978-0-12-819755-4.00010-2

eMDSC	early-stage MDSC
GBM	glioblastoma
G-CSF	granulocyte colony-stimulating factor
GM-CSF	granulocyte-macrophage colony-stimulating factor
IDH	isocitrate dehydrogenase
IDH-mut	mutant isocitrate dehydrogenase
IL-4Rα	IL-4 receptor-α
LGG	low-grade glioma
MDSC	myeloid-derived suppressor cells
MMP	matrix metalloproteinases
NLR	neutrophil/lymphocyte ratio
PBMC	peripheral blood monocyte cell
pHGG	pediatric high-grade glioma
pLGG	pediatric LGG
PMN-MDSC	polymorphonuclear MDSC
ROS	reactive oxygen species
SPP1	osteopontin
STI1	stress-inducible protein-1
TAMs	tumor-associated macrophages
TCGA	The Cancer Genome Atlas
TIMs	tumor-infiltrating myeloid cells
TME	tumor microenvironment
VEGF	vascular endothelial growth factor

Conflicts of interest

HO has the following COI. Consultant fees: Bristol-Myers Squibb, Agios Pharmaceuticals, Grant/Research support: Agios Pharmaceuticals. Royalties as Inventor of: the H3.3K27M TCR, IL-13Ra2 (345–353:1A9V) peptide, and EGFRvIII-CAR for which an exclusive licensing agreement has been executed with Tmunity, Inc., Stemline, Inc. and Novartis Pharma, respectively.

Introduction

Malignant gliomas, in particular grade IV glioblastoma (GBM), are aggressive and complex tumors, composed of both neoplastic and a variety of nonneoplastic cell types [1]. To gain in-depth understanding of glioma and its intricacies, we need to elucidate the tumor microenvironment (TME) and the various cell types that comprise it. In this chapter, we will focus on the myeloid component of GBM, which represents the largest immune subset subpopulation within these tumors. Understanding these cells and their role in glioma can pave the way for future therapeutics via actionable targets or potential adjuvant treatments.

Without inflammatory or pathological conditions, the vast majority of immune cells that are present in the CNS parenchyma are resident microglia [2]. In pathological conditions, such as cancer, the immune profile changes dramatically. The majority of nonneoplastic cells within gliomas are tumor-associated macrophages (TAMs) and a population of immature myeloid cells known as myeloid-derived suppressor cells (MDSCs) [3]. A number of studies have demonstrated that TAMs provide protumorigenic signals to neoplastic cells and support their expansion and invasion through the secretion of growth factors and cytokines [4]. TAMs originate in the blood as monocytes, are recruited to the TME via cytokine gradients produced by glioma cells, and differentiate into protumorigenic, immunosuppressive macrophages [5, 6].

Unlike TAMs, MDSCs exhibit a block in their differentiation program and exhibit their protumorigenic functions primarily through their potent immunosuppressive mechanisms [7, 8]. MDSCs comprise a diverse population of immature myeloid cells that constitute a small percentage of bone marrow cells in healthy individuals [9]. During homeostasis, these cells ultimately differentiate into macrophages, granulocytes, and dendritic cells. On the other hand, during chronic inflammatory conditions, such as cancer, these immature progenitors expand and are recruited into the circulation and the tumor site. In these conditions, the immature myeloid cells encounter low, but persistent levels of stimulation by proinflammatory molecules which inhibit their final differentiation and result in the accumulation of MDSCs [10]. There are currently three widely accepted human MDSC subpopulations which are defined by the expression of cell surface markers: $CD11b^{+}$ $CD33^{+}$ HLA-$DR^{-/low}$ $CD14^{+}$ (monocytic [M-MDSCs]), $CD11b^{+}$ $CD33^{+}$ HLA-$DR^{-/low}$ $CD15^{+}$ (polymononuclear [PMN-MDSCs]), and $Lineage^{-}$ $CD33^{+}$ HLA-DR^{-} (early-stage [eMDSCs]). The latter are the latest described and least well-understood subpopulation. M-MDSCs and PMN-MDSCs are distributed differently throughout the circulation and the TME, and exert distinct immunosuppressive mechanisms [11]. Several studies have illustrated that glioma MDSCs have the ability to attract T regulatory lymphocytes, and also simultaneously block the activation of tumor-reactive $CD4^{+}$ and $CD8^{+}$ T cells [11]. It is instrumental to elucidate their functions through both preclinical studies and analysis of patient samples [12, 13].

Despite decades of work investigating the role of myeloid cells in the formation and progression of glioma, much remains to be evaluated regarding myeloid functionality. As myeloid cells represent up to half the cells within the tumor tissue, their effects upon and interaction with tumor and stromal cells have been implicated in all stages of tumor development and likely also contribute to therapeutic resistance [14, 15]. This chapter aims to highlight the complexity of this immune population, in particular its heterogeneous nature, and discuss the newest emerging strategies to target tumor-infiltrating myeloid cells.

Neutrophils and dendritic cells in glioblastoma biology

Although this chapter will prioritize the discussions of other myeloid populations, it is necessary to recognize the important contributions of neutrophils and dendritic cells (DCs) to the overall composition of the TME, and those are discussed briefly below.

Tumor-derived cytokines naturally recruit neutrophils to the TME, where they can exert both anti- and protumor effects [16]. This variation is dependent on the level of maturation and their activation state (reviewed in [16]). Immature neutrophils are characterized as immunosuppressive, whereas neutrophils with an "aged" phenotype demonstrate an increased ability to respond to inflammatory stimuli and could potentially lead to antitumor response [17]. The number of both circulating and infiltrative neutrophils is also important as it can correlate with the severity of the disease and its progression (Table 1). Glioma patients exhibit strong neutrophilia due to the overproduction of G-CSF (granulocyte colony-stimulating factor) by the TME and/or use of corticosteroids [22]. This strong production of G-CSF drives bone marrow hematopoiesis away from the lymphocyte lineage and toward the production

TABLE 1 Distribution and phenotype of myeloid cells in GBM patients.

Cell type	Low-grade glioma	Adult GBM	High-grade DIPG
Neutrophils	Number of infiltrative neutrophils associated with tumor grade; activation of neutrophils is an early predictor of tumor growth in GBM; increasing number of neutrophils leads to resistance to angiogenic therapies [16]		Future studies
TAMs + microglia	15%–30% of TME (10%–15% in the parenchyma) Majority are $CD11b^{High}CD45^{low}$: microglia [18]	30%–50% of TME Shift toward M2 phenotype, increase tumor proliferation [18]	↑ Antigen presentation ↓ Inflammatory phenotype [19]
MDSCs	Lower percentage of MDSCs in comparison with high grade (correlation between MDSCs and malignancy) [20]	Higher percentage with PBMCs in comparison with LGG patients and those with benign disease [20]	Future studies
Dendritic Cells	cDC1 phenotype recruited and demonstrates antitumoral properties; cDC2 to be elucidated [21]		Future studies

of granulocytes, leading to patients having a high neutrophil/lymphocyte ratio (NLR). Generally, the higher the NLR value, the poorer the prognosis [23]. In glioma preclinical models, larger number of neutrophils implies the promotion of tumor growth and resistance to angiogenic therapies [24, 25]. However, these cells may also be activated to induce an antitumor response.

Dendritic cells are another myeloid population that is excluded from the healthy central nervous system (CNS) parenchyma, but has been shown to gain the ability to migrate to the brain or spinal cord in the setting of pathological conditions [26, 27]. Although the exact role of DCs in GBM is yet to be elucidated, one suggested function for DCs in this context is the recognition and presentation of tumor-specific antigens to help coordinate and elicit a T-cell-mediated response [21]. There have been several studies that have demonstrated the importance of $CD103^+$ dendritic cells (cross-presenting dendritic cells) in antitumor immunity—$CD103^+$ DC-deficient or depleted mouse models have repeatedly shown an inability to reject transplantable immunogenic tumors (breast cancer models, GEM melanoma models, etc.) or support T-cell-based immunotherapies [28]. Despite the paucity of information about the recruitment of endogenous DCs to the GBM microenvironment, a large number of clinical trials are currently testing vaccines utilizing DCs pulsed with glioma-derived antigens [29]. The results of several large Phase II trials are forthcoming; however, earlier study results revealed this approach is feasible and safe, and has the potential to extend patient survival (reviewed in [29]).

Regional distribution of myeloid cells influences their prevalence and phenotype

Most immune cells develop from hematopoietic progenitors in the bone marrow and are later mobilized into the systemic circulation prior to infiltrating the tumor site and becoming terminally differentiated. Thus, it is unsurprising that certain immature populations can be found in blood samples from glioma patients, and their phenotype and function differ from those displayed by cells in the established tumor. This section discusses the differences observed between MDSCs isolated from the peripheral blood monocyte (PBMC) fraction of patients' samples and those derived from the glioma microenvironment.

Circulating MDSCs

MDSCs are readily detected in blood samples of adult glioma patients [7, 20, 30]. Gielen and colleagues demonstrated significantly increased levels of both $CD14^+$ (M-MDSCs) and $CD15^+$ (PMN-MDSCs) cells in blood samples from patients with brain tumors when compared with samples from healthy donors. This observation, however, extended only to GBM samples, whereas MDSC levels from patients with grade II or III gliomas were elevated only slightly above those in healthy donor samples. Even though both MDSC populations were elevated in GBM samples, quantification revealed that M-MDSCs were more abundant in the peripheral blood, while PMN-MDSCs were the predominant population in the analyzed tumor suspensions [30]. Additional studies by Dubinski et al. included a larger cohort of patients and further refined the phenotype of MDSCs found in GBM patients [11]. The latter reported an increase in the frequency of both $CD14^{hi}CD15^{pos}$ (monocytic) and $CD14^{low}CD15^{pos}$ (PMN) MDSCs in the peripheral blood samples of 52 GBM patients when compared with healthy donors. The majority of the PMN-MDSCs, which the authors isolated from the blood of GBM participants, resembled granulocytes, expressed high levels of CD15, and exhibited high T-cell suppressive capabilities [11]. Flow cytometric staining revealed differences in the expression of the class II molecule HLA-DR among the two MDSC subtypes, where $CD14^{hi}CD15^{pos}$ cells exhibited elevated levels of HLA-DR in GBM samples when compared with healthy controls, while $CD14^{low}CD15^{pos}$ cells remained negative. Surprisingly, the authors detected very low levels of arginase activity from both subsets of freshly isolated blood-derived MDSCs; however, both cell types produced high levels of reactive oxygen species (ROS) in a functional assay. These detailed characterizations of both the phenotypes and functions of MDSCs indicate that these immature myeloid cells represent a highly heterogeneous population that varies with tumor grade and differs from MDSCs found at the tumor site.

The studies discussed above suggest that developing a deeper understanding of the dynamics and defining characteristics of these populations will help identify potential biomarkers of disease progression and response to therapy. A recent publication by Alban et al. extensively profiled the accumulation and phenotypes of circulating MDSCs in patients diagnosed with GBM and other brain malignancies. When compared with samples from patients with nonmalignant brain tumors and low-grade gliomas (LGG), MDSCs from GBM patients (defined as bulk $HLA\text{-}DR^{-/lo}$, $CD33^+$, $CD11b^+$) represented a higher percentage of total PBMCs, while there was no significant difference between the levels found in grade I/II or III and benign samples. These observations led the authors to propose that the percentage of MDSCs found in peripheral blood might be a marker of brain tumor malignancy [20]. Multivariate analyses additionally indicated that in the cohort of patients examined, tumor grade was the factor most significantly associated with MDSC levels in patients' blood. The authors evaluated blood samples longitudinally over the course of surgery and at regular intervals post tumor resection until the patients succumbed to the disease. This allowed them to characterize the immune profile changes in six GBM patients. Utilizing a 25-marker mass cytometry panel, Alban et al. revealed that patients who had a good prognosis (survival over 20 months) displayed a decreased MDSC population over time, while those with poor prognosis had increasing levels of MDSCs, suggesting that measurements of circulating MDSCs may indicate disease progression. The study also identified CD11c^{+} DCs and increased levels of HLA-DR to be associated with a good prognosis in GBM patients. In the same study, the authors compared the immune profiles of patients with LGG and GBM at baseline, and noted that levels of DCs and natural killer cells were higher in LGG patients, leading them to hypothesize that LGG patients are initially better primed to mount an effective antitumor immune response [20]. These findings strongly suggest that longitudinal characterization of circulating myeloid cells may provide critical information about disease progression and overall survival in brain tumor patients.

MDSCs within the tumor microenvironment

Inside the TME, tumor-associated macrophages and microglia (reviewed below) in addition to tumor-infiltrating immature myeloid cells, such as MDSCS, exert a profound immunosuppressive effect and likely limit the efficacy of immunotherapies tested in the clinic thus far [31]. Often distinguishing precisely among these populations is arduous given that they are all derived from myeloid progenitors and represent a spectrum of differentiation states.

Immature myeloid cells within the TME remain the least well understood. Unlike patients' peripheral blood samples where both major subtypes of MDSCs are elevated, PMN-MDSCs represent the majority of MDSCs in tumor samples [11, 30]. Variations as to the precise levels of each differ among studies, likely due to the size of patient cohorts and sensitivity of flow cytometric analyses available at the time of publication. As discussed in the previous section, Dubinski and colleagues examined MDSCs in the tumors of 52 GBM patients and reported significant differences in their phenotype and function when compared with the same populations found in circulation [11]. Monocytic MDCSs displayed progressively higher levels of HLA-DR and CD124 (IL4Rα) when compared with expression in the PBMCs of healthy donors and GBM patients, while their levels of CD16 diminished [11]. In contrast, the expression of CD124 and CD16 remained constant in most PMN-MDSC subsets, indicating that the two different MDSC populations exhibit separate activation states. In a later study, Alban et al. profiled the levels of MDSCs in patients with primary and recurrent GBM and found that patients with high levels of MDSCs in recurrent samples had a worse overall survival, decreased time between first and second surgery, and worse survival after second surgery [20]. In the same analysis, levels of MDSCs in primary tumors were not predictive of patient outcome. The authors concurrently evaluated whether this predictive potential was specific to MDSCs or whether it extended to other tumor-associated myeloid cells. Interestingly, high levels of bulk myeloid cells (defined as CD33^{+} cells) were associated with increased patient survival, underscoring the heterogeneous nature of myeloid cell populations within GBM tumors [20].

Profiling studies, such as the one described above, provide a valuable insight into the prevalence, distribution, and significance to patient outcomes. Elucidating the biological impact of myeloid cells in the TME, however, benefits greatly from preclinical animal studies where hypotheses can be tested and therapeutic treatments can be evaluated. One such study by Antonios et al. investigated the contribution of tumor-infiltrating myeloid cells (TIMs, defined as CD163$^+$ myelomonocytic cells, likely including MDSCs) to local immune suppression via the PD-1/PD-L1 axis [13]. The authors demonstrated that while both tumor and myeloid cells can up-regulate PD-L1 expression in vitro, PD-L1 expression by TIMs within an in vivo animal model and within patient-derived cells was significantly higher than the levels detected on tumor cells. These levels further increased following a vaccine-based treatment. Depletion or modulation of TIMs' activation state combined with PD-1 blockade proved synergistic and was able to reverse the immune suppression mounted in response to vaccine treatment resulting in dramatically increased survival in a GL261-based animal model [13]. These results further indicate that myeloid cells within the TME may display significantly different characteristics when compared with circulating cells, likely due to the effects of locally produced tumor and stromal-derived factors. While Gielen et al. did not observe detectable levels of PD-L1 in circulating MDSCs, the data from Antonios and colleagues provide strong evidence that PD-L1 expression by myeloid cells within the TME is a critical component of the glioma immune suppression mechanism [30].

Another element that MDSCs encounter at the tumor site, and not in circulation, is hypoxia. Exposure to hypoxia and additional stress signals in the TME has been shown to alter the metabolic pathways utilized by all myeloid cells, including MDSCs (reviewed in [8]). The metabolic enzyme arginase I (ARG1) is often used as the key functional marker of MDSCs as a plethora of studies have demonstrated that up-regulation of ARG1 activity in MDSCs can deplete the levels of the amino acid L-arginine and therefore alter the ability of cytotoxic lymphocytes to proliferate and become activated. Our group demonstrated that in both mouse and human GBM, IL-4 receptor-α (IL-4Rα) is up-regulated on glioma-infiltrating MDSCs, but not in myeloid cells in the periphery or in normal brain tissue, where it mediates IL-13-induced production of arginase [32]. We found that granulocyte-macrophage colony-stimulating factor (GM-CSF) is up-regulated in both human and mouse glioma environments compared with normal brain or peripheral blood samples, and plays a key role for the induction of IL-4Rα expression on myeloid cells. Taken together, our study demonstrated a novel GM-CSF-induced immunosuppression mechanism in the glioma environment via IL-4Rα and arginase [32]. The precise downstream effect of ARG1 signaling on the metabolome of MDSCs found in GBM, however, remains undefined and is an emerging area of research in the field.

Delineating the differences between circulating and tumor-infiltrating immune cells carries the potential of elucidating the major pathways affecting the immune cells as they extravasate from the peripheral blood and enter the TME. These pathways can provide novel therapeutic targets, in addition to those already being tested in the clinic, such as PD-1/PD-L1 blockade and inhibition of the survival cytokine colony-stimulating factor 1 (CSF-1). For instance, Chang et al. demonstrated that in both preclinical mouse models and within GBM patient samples, the chemokine CCL2 is highly produced by myeloid cells in the TME and acts to recruit immunosuppressive cells such as regulatory T cells and MDSCs. Blockade of CCL2 and its receptor CCR4 in vivo, using both genetic manipulation and a small-molecule antagonist, significantly diminished the levels of suppressive cells in the TME and prolonged the survival of tumor-bearing animals [33]. In addition to new therapeutic agents, however, recent studies underscore the need for combinatorial treatment approaches as single treatments almost inevitably result in immunosuppressive feedback mechanisms and tumor recurrence [13, 14, 20]. Gathering MDSC samples from circulating PBMCs is noninvasive and offers the possibility of longitudinal monitoring which is indispensable in a population of patients where multiple tumor biopsies are rare due to the tumors' anatomical location and the aggressive nature of the disease, thus providing a new avenue for treatment and disease progression monitoring.

Intratumoral heterogeneity of macrophages/microglia

The immune microenvironment of grade II to grade IV gliomas contains a large proportion of brain-resident microglia and bone-marrow derived TAMs. Glioma cells reprogram TAMs to support their own growth and progression through the release of a multitude of molecules such as soluble lipids, cytokines, chemokines, and extracellular vesicles [33–35]. The composition and activation state of TAMs in the TME varies with molecular subtype of glioma. A single-cell transcriptomics study revealed that the population of TAMs in glioma can progress from strictly microglia in the early phases to a mixture of microglia and infiltrating monocytes and macrophages in the late phases of tumor growth [36]. Bone marrow-derived TAMs but not microglia are the dominant myeloid population in GBM compared with low-grade glioma [36]. Moreover, the degree of bone-marrow derived TAMs infiltrating the TME is higher in mesenchymal GBM compared with the classical and proneural subtype [36]. On the other hand, infiltration of microglial TAMs is high in low-grade astrocytomas as opposed to oligodendrogliomas and GBMs. Interrogation of the Ivy Glioblastoma Atlas Project dataset for TAM gene signature revealed enrichment of the microglial gene signature in samples obtained from the leading edge of invading gliomas and in infiltrated white matter [36]. The gene signature for bone marrow-derived TAMs was high in regions of hyperplastic blood vessels, microvascular proliferation and in peri-necrotic regions. Flow cytometric analysis of the tumor core versus margins in grade IV GBMs revealed the accumulation of bone marrow-derived macrophages in the center of the neoplastic lesion [37].

The innate immune microenvironment in untreated isocitrate dehydrogenase (IDH)-mutant GBM (IDH-mut) is markedly different from the one found in wild-type GBM. The percentage of microglia and bone-marrow derived macrophages is strikingly lower in IDH-mut glioma as compared with IDH-wildtype GBM. Moreover, the myeloid population in IDH-mut glioma have a proinflammatory gene signature [38].

Cellular heterogeneity of the myeloid compartment in pediatric CNS tumors has not been studied as much as that in adult gliomas. A recent study evaluated immune infiltrate in three types of pediatric CNS tumors, namely, pLGG (grade I pilocytic astrocytoma), pHGG (anaplastic astrocytoma and GBM), and diffuse intrinsic pontine gliomas (DIPG) [39]. Both pLGG and pHGG were shown to have considerable accumulation of $CD163^+$ macrophages, while the percentage of $CD45^+CD11b^+$ cells in DIPG tumors was similar to that in nontumor cortical tissue. The authors also indicated that DIPG-derived myeloid cells do not overexpress immunosuppressive factors, such as PD-L1, compared with those derived from nontumor cortical tissue. Another recent study quantified the percentage of $CD45^+$ cells which are also $CD11b^+$ in DIPG and adult GBM, and found a significant difference between the two patient cohorts ($CD11b^+$ among $CD45^+$ cells, DIPG: 94.80% ± 0.92% vs adult GBM: 70.34% ± 7.20%, $P < .005$) [19]. Lin et al. also reported that myeloid cells in DIPG express less inflammatory cytokines compared with those in adult GBM [19] (Table 1). There is a lack of data on spatial and temporal intratumor heterogeneity for DIPG cases due to their unique anatomical location, rendering repeated or extensive sampling of tumor tissue particularly challenging.

Taken together, these studies highlight the phenotypical and functional diversity of glioma-associated macrophage/microglia.

Molecular mechanisms driving the induction, recruitment, and function of glioma-associated myeloid cells

Similar to other solid tumors, glioma cells secrete a variety of chemokines, thus creating a gradient that recruits myeloid cells from the blood. Once within the tumor microenvironment, TAMs, together with the resident microglia, encounter additional secreted factors that shape their protumorigenic phenotype. The following section discusses the molecular pathways that dictate the complex interactions between tumor-associated myeloid cells and the tumor compartment.

Interaction between bone marrow-derived macrophages and glioma

Glioblastoma cells release cytokines, chemokines, lipids, metabolites, and extracellular vesicles that modulate the activation and function of TAMs. Several studies have explored the mechanisms underlying the recruitment of bone marrow-derived macrophages to the tumor site. A recent study reported osteopontin (SPP1) as a potent chemokine for macrophages, and its blockade significantly impaired the ability of glioma cells to recruit macrophages [40]. Glioma stem cells (GSCs) in the perivascular niches secrete periostin that additionally functions as a chemoattractant, recruiting monocytes from the peripheral blood [41]. Furthermore, it has been demonstrated that the hypoxic environment inside the TME also contributes to the up-regulation of periostin expression in GSCs [42]. In PTEN-deficient GBMs, glioma cells secrete lysyl-oxidase that functions as macrophage-chemoattractant via activation of the beta1 integrin-PYK2 pathway in macrophages [43]. In addition, chemokines and cytokines—CCL2, CSF1, VEGF, SDF1, GDNF, CCL7, and ATP—have also been shown to mediate TAM infiltration [33, 44–46]. Inactivation of NF1 in mesenchymal GBM enhances TAM infiltration [47]. It has been reported that mesenchymal GBM exhibits higher TAM infiltration compared with other molecular subtypes. Oncometabolites in the tumor microenvironment also act on TAMs to enhance their protumor functions. Takenaka et al. recently reported that kynurenine, an oncometabolite secreted by GBM cells, induced the activation of the aryl hydrocarbon receptor (AHR) pathway in macrophages [48]. Activation of AHR impacts macrophage function, polarization, and recruitment. This study demonstrated that AHR induces expression of the transcription factor KLF4 and suppresses NF-kB activation via a SOCS2-TRAF6 mechanism. In GBM patients, the authors showed that high expression of AHR is associated with poor prognosis [48].

A recent study uncovered the mechanism by which TAMs maintain GSCs and promote malignant growth [49]. Shi et al. demonstrated that TAMs secrete the cytokine pleiotrophin which signals through the receptor PTPRZ1 on GSCs and maintains their tumorigenic potential. Analysis of The Cancer Genome Atlas (TCGA) dataset revealed the association between higher pleiotrophin and poorer progression-free survival for GBM patients [49].

Recent studies have demonstrated that the macrophage differentiation in the TME is much more complex than the classical binary model of macrophage polarization: M1 (tumor suppressive) and M2 (tumor promoting) [36, 50, 51]. Single-cell transcriptional analysis showed that GBM-infiltrating TAMs co-expressed transcripts linked to both pro- (M1) and antiinflammatory (M2) phenotypes [36]. This noncanonical phenotype of TAMs has been validated at the protein level and suggests that the transcriptional program of TAMs includes multiple modules to support their pathogenic role in GBM progression.

The role of TAMs in promoting tumor proliferation, invasiveness, and neoangiogenesis has been well elucidated by several studies. One of the outcomes of the interaction between GBM cells and TAMs is tissue remodeling, which enables GBM cells to invade the brain parenchyma. Matrix metalloproteinases (MMPs) are crucial in tissue remodeling and wound repair [52]. TAM-derived TGFβ1 promotes MMP9 induction in GSCs, which enhances their invasiveness [53]. In PTEN-null GBM mouse models, infiltrating macrophages secrete SPP1 which promotes glioma cell survival and angiogenesis [43]. Accordingly, in the TCGA GBM patient dataset, higher SPP1 expression levels positively correlate with higher angiogenesis and lower patient survival.

Owing to rapid growth, gliomas are in constant need of neovascularization and as a result release the angiogenic factors epidermal growth factor (EGF) and vascular endothelial growth factor (VEGF) [54]. In addition, TAMs accumulate around blood vessels and also produce angiogenic factors such as VEGF and CXCL12 [55, 56]. Recently, the extracellular adenosine deaminase protein cat eye syndrome critical region protein 1 (CECR1) was shown to contribute to the proangiogenic function of glioma TAMs. Analysis of human GBM samples by both immunohistochemical staining and TCGA dataset mining indicated that levels of CECR1 positively correlate with microvascular density. In glioblastoma, CECR1 produced by M2-like TAMs regulates the cross talk between macrophages and pericytes via paracrine PDGFB-PDGFRβ signaling, promoting pericyte recruitment and migration, and tumor angiogenesis [57].

Overall, the interaction between glioma cells and macrophages is bidirectional and multifactorial. There are multiple ligand-receptor interactions that ultimately determine TAMs' effect on tumor growth. Improved understanding of these interactions could lead to new actionable targets for glioma treatment.

Interaction between microglia and glioma

Rapid proliferation and dissemination of glioma cells in the brain parenchyma are achieved by several mechanisms and involve interplay between microglia and glioma cells. Glioma cells repurpose microglia such that they no longer perform immune surveillance but instead are protumorigenic. A study for Hu et al. demonstrated that glioma cells secrete the extracellular matrix protein versican that signals via toll-like receptor 2, present on microglia, and induces the release of MMP14. MMP14 is then available to convert the glioma-secreted precursor of MMP2 into its active state, thereby facilitating the invasion of tumor cells into the brain parenchyma [58]. EGF has also been implicated in gliomagenesis. Microglia-derived EGF increases tumor invasion via the activation of its receptors on glioma cells [59]. Microglia surrounding the tumor additionally express ARG1, which has been shown to promote tumor cell proliferation [60]. Similarly, the release of stress-inducible protein-1 (STI1) by microglia induces tumor growth and invasion through MMP9 production [61]. GBM-derived Wnt3a further induces a protumorigenic phenotype in microglia through up-regulation of ARG1, STI1, IL-10, and a decrease in IL-1β gene expression [62]. Furthermore, in vitro studies have shown that CSF-1 secreted by GBM cells can induce insulin-like-growth factor-binding protein in microglia, which can stimulate angiogenesis [63].

Current therapeutic strategies and future directions

The findings discussed thus far highlight the diverse and complex landscape of myeloid cells in patients with GBM. Targeting this subset of immunosuppressive cells often addresses one of the three main steps of myeloid cell development and function. These steps include (1) inhibition of trafficking and recruitment of the myeloid cells from the circulation into the TME, (2) repolarization of TAMs toward more proinflammatory, antitumor phenotypes, and (3) promoting the differentiation of MDSCs or inhibiting their survival and accumulation (summarized in Fig. 1). Recent work by Ding et al. extensively reviewed these mechanisms and the currently available biologics and small-molecule inhibitors that target these pathways [64]. In this concluding section, we highlight the newest discoveries in the field of myeloid cell-based therapies for solid tumors, which were not available at the time of publication of the work by Ding and colleagues.

One of the defining phenotypic markers of MDSCs is their expression of CD33. This molecule is also known as Siglec-3 and is a member of the sialic acid-binding immunoglobulin-like lectin (Siglec) family. Members of this family of receptors are expressed primarily by immune cells and are broadly detected in most subpopulations. As their intracellular domains contain an immunoreceptor tyrosine-based inhibitory motif, binding to their widely expressed ligands can result in downregulation of immune signaling and may potentially contribute to the immunosuppressive landscape observed in tumors [65]. Santegoets et al. profiled the expression of different Siglec family members on MDSCs found in the blood and tumor tissues of glioma patients and compared them with those observed on MDSCs isolated from healthy donors [66]. They reported that both M- and PMN-MDSCs express multiple Siglec receptors, the exact profiles differing between the two MDSC populations and similar levels of expression were found in healthy donors as well as glioma patients. Additionally, the authors confirmed the Siglec expression profile on PMN-MDSCs within patient-derived glioma tissue. The study also revealed the expression of possible Siglec ligands in GBM cell lines and in GBM patient samples. These data suggest that Siglec-sialic acid interactions may be contributing to the phenotype and function of myeloid cells in glioma patients. Further investigation into these interactions may reveal novel molecular pathways that can guide the immune microenvironment and provide previously unexplored therapeutic targets.

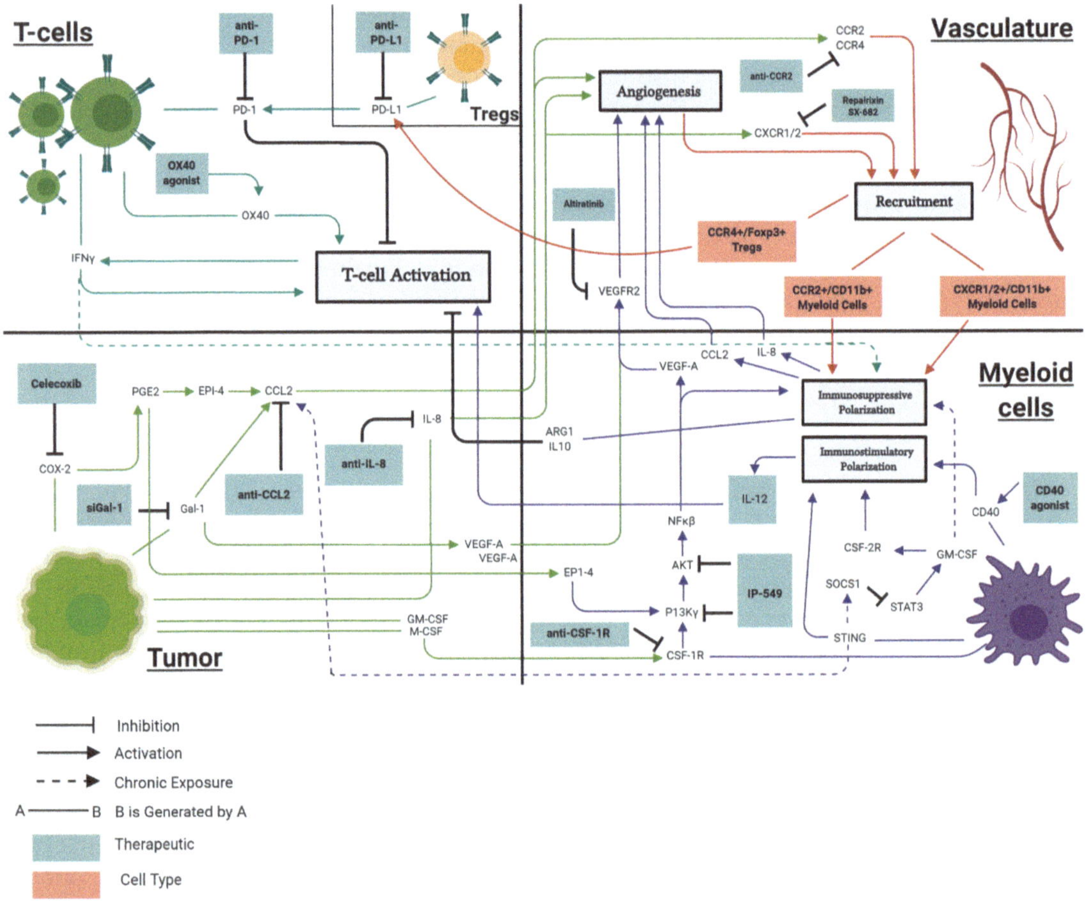

FIG. 1 A schematic summary of the complex interactions between cells in GBM TME and potential steps for therapeutic intervention. *Adapted and updated from Ding AS, Routkevitch D, Jackson C, Lim M. Targeting myeloid cells in combination treatments for glioma and other tumors. Front Immunol 2019;10:1715.*

A recent study by Schmid et al. reported on a novel function of another well-described myeloid cell surface marker, CD11b [67]. The authors' data indicated that CD11b may play an important role in regulating the levels of proinflammatory responses in macrophages and loss of its expression or function promotes up-regulation of immunosuppressive genes and in vivo recruitment of regulatory T cells. Expression of the miRNA Let7a and c-Myc was implicated as the downstream effectors of CD11b activation. Treating tumor-bearing animals with the CD11b agonist LA1 resulted in altered TAM polarization, higher levels of intratumoral CD8^{+} T cells, and prolonged animal survival. The study suggested broad implications stemming from CD11b signaling including changes in innate and adaptive immune responses and tumor vascularization [67]. A study conducted by Panni et al. in a different preclinical model broadly supported the notion that CD11b agonists alter macrophage polarization, increase cytotoxic T-cell responses, and render tumors susceptible to additional immunotherapeutic treatments [67]. Despite potential differences in specific mechanism of action, both studies demonstrated a novel pathway regulating the function of myeloid cells in solid tumors with remarkable antitumor responses in preclinical models. The small molecule tested by Panni et al., ADH-503, is soon slated to begin phase I single-agent clinical trials [67]. The data presented by these two groups are extremely promising; however, previous solid tumor clinical trials targeting single immune cell pathways have rarely been successful and the greatest patient benefit is likely to be observed in combination therapies, such as the addition of immune checkpoint inhibitors to the treatment of animals with ADH-503 reported in the preclinical setting by Panni et al. An extensive summary of the current clinical trials targeting myeloid cells in combination with other cancer therapies is provided by Ding and colleagues, and the results of many of them are still forthcoming [64].

It is evident that breakthroughs of immunotherapy for glioma patients will require effective targeting of myeloid cells that heavily infiltrate these deadly tumors. Moving forward, we hope that recent and ongoing discoveries on the complex interplays between glioma and myeloid cells will allow us to design more rational combination strategies in near future clinical trials.

References

[1] Patel AP, Tirosh I, Trombetta JJ, Shalek AK, Gillespie SM, Wakimoto H, Cahill DP, Nahed BV, Curry WT, Martuza RL, Louis DN, Rozenblatt-Rosen O, Suva ML, Regev A, Bernstein BE. Single-cell RNA-seq highlights intratumoral heterogeneity in primary glioblastoma. Science 2014;344:1396–401.

[2] Grauer OM, Wesseling P, Adema GJ. Immunotherapy of diffuse gliomas: biological background, current status and future developments. Brain Pathol 2009;19:674–93.

[3] Hussain SF, Yang D, Suki D, Aldape K, Grimm E, Heimberger AB. The role of human glioma-infiltrating microglia/macrophages in mediating antitumor immune responses. Neuro-Oncology 2006;8:261–79.

[4] Li Q, Barres BA. Microglia and macrophages in brain homeostasis and disease. Nat Rev Immunol 2018;18:225–42.

[5] Parney IF, Waldron JS, Parsa AT. Flow cytometry and in vitro analysis of human glioma-associated macrophages. Laboratory investigation. J Neurosurg 2009;110:572–82.

[6] Wu A, Wei J, Kong LY, Wang Y, Priebe W, Qiao W, Sawaya R, Heimberger AB. Glioma cancer stem cells induce immunosuppressive macrophages/microglia. Neuro-Oncology 2010;12:1113–25.

[7] Raychaudhuri B, Rayman P, Ireland J, Ko J, Rini B, Borden EC, Garcia J, Vogelbaum MA, Finke J. Myeloid-derived suppressor cell accumulation and function in patients with newly diagnosed glioblastoma. Neuro-Oncology 2011;13:591–9.

[8] Kumar V, Patel S, Tcyganov E, Gabrilovich DI. The nature of myeloid-derived suppressor cells in the tumor microenvironment. Trends Immunol 2016;37:208–20.

[9] Budhwar S, Verma P, Verma R, Rai S, Singh K. The Yin and Yang of myeloid derived suppressor cells. Front Immunol 2018;9:2776.

[10] Trikha P, Carson 3rd WE. Signaling pathways involved in MDSC regulation. Biochim Biophys Acta 2014;1846:55–65.

[11] Dubinski D, Wolfer J, Hasselblatt M, Schneider-Hohendorf T, Bogdahn U, Stummer W, Wiendl H, Grauer OM. CD4+ T effector memory cell dysfunction is associated with the accumulation of granulocytic myeloid-derived suppressor cells in glioblastoma patients. Neuro-Oncology 2016;18:807–18.

[12] Gielen PR, Schulte BM, Kers-Rebel ED, Verrijp K, Bossman SA, Ter Laan M, Wesseling P, Adema GJ. Elevated levels of polymorphonuclear myeloid-derived suppressor cells in patients with glioblastoma highly express S100A8/9 and arginase and suppress T cell function. Neuro-Oncology 2016;18:1253–64.

[13] Antonios JP, Soto H, Everson RG, Moughon D, Orpilla JR, Shin NP, Sedighim S, Treger J, Odesa S, Tucker A, Yong WH, Li G, Cloughesy TF, Liau LM, Prins RM. Immunosuppressive tumor-infiltrating myeloid cells

mediate adaptive immune resistance via a PD-1/PD-L1 mechanism in glioblastoma. Neuro-Oncology 2017;19:796–807.

[14] Quail DF, Bowman RL, Akkari L, Quick ML, Schuhmacher AJ, Huse JT, Holland EC, Sutton JC, Joyce JA. The tumor microenvironment underlies acquired resistance to CSF-1R inhibition in gliomas. Science 2016;352(6288), aad3018.

[15] Gubin MM, Esaulova E, Ward JP, Malkova ON, Runci D, Wong P, Noguchi T, Arthur CD, Meng W, Alspach E, Medrano RFV, Fronick C, Fehlings M, Newell EW, Fulton RS, Sheehan KCF, Oh ST, Schreiber RD, Artyomov MN. High-dimensional analysis delineates myeloid and lymphoid compartment remodeling during successful immune-checkpoint cancer therapy. Cell 2018;175:1014–30. e19.

[16] Massara M, Persico P, Bonavita O, Mollica Poeta V, Locati M, Simonelli M, Bonecchi R. Neutrophils in gliomas. Front Immunol 2017;8:1349.

[17] Adrover JM, Nicolas-Avila JA, Hidalgo A. Aging: a temporal dimension for neutrophils. Trends Immunol 2016;37:334–45.

[18] Hambardzumyan D, Gutmann DH, Kettenmann H. The role of microglia and macrophages in glioma maintenance and progression. Nat Neurosci 2016;19:20–7.

[19] Lin GL, Nagaraja S, Filbin MG, Suva ML, Vogel H, Monje M. Non-inflammatory tumor microenvironment of diffuse intrinsic pontine glioma. Acta Neuropathol Commun 2018;6:51.

[20] Alban TJ, Alvarado AG, Sorensen MD, Bayik D, Volovetz J, Serbinowski E, Mulkearns-Hubert EE, Sinyuk M, Hale JS, Onzi GR, McGraw M, Huang P, Grabowski MM, Wathen CA, Ahluwalia MS, Radivoyevitch T, Kornblum HI, Kristensen BW, Vogelbaum MA, Lathia JD. Global immune fingerprinting in glioblastoma patient peripheral blood reveals immune-suppression signatures associated with prognosis. JCI Insight 2018;3.

[21] Srivastava S, Jackson C, Kim T, Choi J, Lim M. A characterization of dendritic cells and their role in immunotherapy in glioblastoma: from preclinical studies to clinical trials. Cancers (Basel) 2019;11.

[22] Nitta T, Sato K, Allegretta M, Brocke S, Lim M, Mitchell DJ, Steinman L. Expression of granulocyte colony stimulating factor and granulocyte-macrophage colony stimulating factor genes in human astrocytoma cell lines and in glioma specimens. Brain Res 1992;571:19–25.

[23] Wiencke JK, Koestler DC, Salas LA, Wiemels JL, Roy RP, Hansen HM, Rice T, McCoy LS, Bracci PM, Molinaro AM, Kelsey KT, Wrensch MR, Christensen BC. Immunomethylomic approach to explore the blood neutrophil lymphocyte ratio (NLR) in glioma survival. Clin Epigenetics 2017;9:10.

[24] Achyut BR, Shankar A, Iskander AS, Ara R, Angara K, Zeng P, Knight RA, Scicli AG, Arbab AS. Bone marrow derived myeloid cells orchestrate antiangiogenic resistance in glioblastoma through coordinated molecular networks. Cancer Lett 2015;369:416–26.

[25] Liang J, Piao Y, Holmes L, Fuller GN, Henry V, Tiao N, de Groot JF. Neutrophils promote the malignant glioma phenotype through S100A4. Clin Cancer Res 2014;20:187–98.

[26] Jain P, Coisne C, Enzmann G, Rottapel R, Engelhardt B. Alpha4beta1 integrin mediates the recruitment of immature dendritic cells across the blood-brain barrier during experimental autoimmune encephalomyelitis. J Immunol 2010;184:7196–206.

[27] John B, Ricart B, Tait Wojno ED, Harris TH, Randall LM, Christian DA, Gregg B, De Almeida DM, Weninger W, Hammer DA, Hunter CA. Analysis of behavior and trafficking of dendritic cells within the brain during toxoplasmic encephalitis. PLoS Pathog 2011;7, e1002246.

[28] Broz ML, Binnewies M, Boldajipour B, Nelson AE, Pollack JL, Erle DJ, Barczak A, Rosenblum MD, Daud A, Barber DL, Amigorena S, Van't Veer LJ, Sperling AI, Wolf DM, Krummel MF. Dissecting the tumor myeloid compartment reveals rare activating antigen-presenting cells critical for T cell immunity. Cancer Cell 2014;26:638–52.

[29] Eagles ME, Nassiri F, Badhiwala JH, Suppiah S, Almenawer SA, Zadeh G, Aldape KD. Dendritic cell vaccines for high-grade gliomas. Ther Clin Risk Manag 2018;14:1299–313.

[30] Gielen PR, Schulte BM, Kers-Rebel ED, Verrijp K, Petersen-Baltussen HM, ter Laan M, Wesseling P, Adema GJ. Increase in both CD14-positive and CD15-positive myeloid-derived suppressor cell subpopulations in the blood of patients with glioma but predominance of CD15-positive myeloid-derived suppressor cells in glioma tissue. J Neuropathol Exp Neurol 2015;74:390–400.

[31] Kamran N, Chandran M, Lowenstein PR, Castro MG. Immature myeloid cells in the tumor microenvironment: implications for immunotherapy. Clin Immunol 2018;189:34–42.

[32] Kohanbash G, McKaveney K, Sakaki M, Ueda R, Mintz AH, Amankulor N, Fujita M, Ohlfest JR, Okada H. GM-CSF promotes the immunosuppressive activity of glioma-infiltrating myeloid cells through interleukin-4 receptor-alpha. Cancer Res 2013;73:6413–23.

[33] Chang AL, Miska J, Wainwright DA, Dey M, Rivetta CV, Yu D, Kanojia D, Pituch KC, Qiao J, Pytel P, Han Y, Wu M, Zhang L, Horbinski CM, Ahmed AU, Lesniak MS. CCL2 produced by the glioma microenvironment is essential for the recruitment of regulatory T cells and myeloid-derived suppressor cells. Cancer Res 2016;76:5671–82.

[34] Loh JK, Hwang SL, Lieu AS, Huang TY, Howng SL. The alteration of prostaglandin E2 levels in patients with brain tumors before and after tumor removal. J Neuro-Oncol 2002;57:147–50.

[35] Kjellman C, Olofsson SP, Hansson O, Von Schantz T, Lindvall M, Nilsson I, Salford LG, Sjogren HO, Widegren B. Expression of TGF-beta isoforms, TGF-beta receptors, and SMAD molecules at different stages of human glioma. Int J Cancer 2000;89:251–8.

[36] Muller S, Kohanbash G, Liu SJ, Alvarado B, Carrera D, Bhaduri A, Watchmaker PB, Yagnik G, Di Lullo E, Malatesta M, Amankulor NM, Kriegstein AR, Lim DA, Aghi M, Okada H, Diaz A. Single-cell profiling of human gliomas reveals macrophage ontogeny as a basis for regional differences in macrophage activation in the tumor microenvironment. Genome Biol 2017;18:234.

[37] Pinton L, Masetto E, Vettore M, Solito S, Magri S, D'Andolfi M, Del Bianco P, Lollo G, Benoit JP, Okada H, Diaz A, Della Puppa A, Mandruzzato S. The immune suppressive microenvironment of human gliomas depends on the accumulation of bone marrow-derived macrophages in the center of the lesion. J Immunother Cancer 2019;7:58.

[38] Poon CC, Gordon PMK, Liu K, Yang R, Sarkar S, Mirzaei R, Ahmad ST, Hughes ML, Yong VW, Kelly JJP. Differential microglia and macrophage profiles in human IDH-mutant and -wild type glioblastoma. Oncotarget 2019;10:3129–43.

[39] Lieberman NAP, DeGolier K, Kovar HM, Davis A, Hoglund V, Stevens J, Winter C, Deutsch G, Furlan SN, Vitanza NA, Leary SES, Crane CA. Characterization of the immune microenvironment of diffuse intrinsic pontine glioma: implications for development of immunotherapy. Neuro-Oncology 2019;21:83–94.

[40] Wei J, Marisetty A, Schrand B, Gabrusiewicz K, Hashimoto Y, Ott M, Grami Z, Kong LY, Ling X, Caruso H, Zhou S, Wang YA, Fuller GN, Huse J, Gilboa E, Kang N, Huang X, Verhaak R, Li S, Heimberger AB. Osteopontin mediates glioblastoma-associated macrophage infiltration and is a potential therapeutic target. J Clin Invest 2019;129:137–49.

[41] Zhou W, Ke SQ, Huang Z, Flavahan W, Fang X, Paul J, Wu L, Sloan AE, McLendon RE, Li X, Rich JN, Bao S. Periostin secreted by glioblastoma stem cells recruits M2 tumour-associated macrophages and promotes malignant growth. Nat Cell Biol 2015;17:170–82.

[42] Guo X, Xue H, Shao Q, Wang J, Guo X, Chen X, Zhang J, Xu S, Li T, Zhang P, Gao X, Qiu W, Liu Q, Li G. Hypoxia promotes glioma-associated macrophage infiltration via periostin and subsequent M2 polarization by upregulating TGF-beta and M-CSFR. Oncotarget 2016;7:80521–42.

[43] Chen P, Zhao D, Li J, Liang X, Li J, Chang A, Henry VK, Lan Z, Spring DJ, Rao G, Wang YA, DePinho RA. Symbiotic macrophage-glioma cell interactions reveal synthetic lethality in PTEN-null glioma. Cancer Cell 2019;35:868–84. e6.

[44] Zhu X, Fujita M, Snyder LA, Okada H. Systemic delivery of neutralizing antibody targeting CCL2 for glioma therapy. J Neuro-Oncol 2011;104:83–92.

[45] Ku MC, Wolf SA, Respondek D, Matyash V, Pohlmann A, Waiczies S, Waiczies H, Niendorf T, Synowitz M, Glass R, Kettenmann H. GDNF mediates glioblastoma-induced microglia attraction but not astrogliosis. Acta Neuropathol 2013;125:609–20.

[46] Okada M, Saio M, Kito Y, Ohe N, Yano H, Yoshimura S, Iwama T, Takami T. Tumor-associated macrophage/microglia infiltration in human gliomas is correlated with MCP-3, but not MCP-1. Int J Oncol 2009;34:1621–7.

[47] Wang Q, Hu B, Hu X, Kim H, Squatrito M, Scarpace L, deCarvalho AC, Lyu S, Li P, Li Y, Barthel F, Cho HJ, Lin YH, Satani N, Martinez-Ledesma E, Zheng S, Chang E, Sauve CG, Olar A, Lan ZD, Finocchiaro G, Phillips JJ, Berger MS, Gabrusiewicz KR, Wang G, Eskilsson E, Hu J, Mikkelsen T, DePinho RA, Muller F, Heimberger AB, Sulman EP, Nam DH, Verhaak RGW. Tumor evolution of glioma-intrinsic gene expression subtypes associates with immunological changes in the microenvironment. Cancer Cell 2017;32:42–56. e6.

[48] Takenaka MC, Gabriely G, Rothhammer V, Mascanfroni ID, Wheeler MA, Chao CC, Gutierrez-Vazquez C, Kenison J, Tjon EC, Barroso A, Vandeventer T, de Lima KA, Rothweiler S, Mayo L, Ghannam S, Zandee S, Healy L, Sherr D, Farez MF, Prat A, Antel J, Reardon DA, Zhang H, Robson SC, Getz G, Weiner HL, Quintana FJ. Control of tumor-associated macrophages and T cells in glioblastoma via AHR and CD39. Nat Neurosci 2019;22:729–40.

[49] Shi Y, Ping YF, Zhou W, He ZC, Chen C, Bian BS, Zhang L, Chen L, Lan X, Zhang XC, Zhou K, Liu Q, Long H, Fu TW, Zhang XN, Cao MF, Huang Z, Fang X, Wang X, Feng H, Yao XH, Yu SC, Cui YH, Zhang X, Rich JN, Bao S, Bian XW. Tumour-associated macrophages secrete pleiotrophin to promote PTPRZ1 signalling in glioblastoma stem cells for tumour growth. Nat Commun 2017;8:15080.

[50] Szulzewsky F, Pelz A, Feng X, Synowitz M, Markovic D, Langmann T, Holtman IR, Wang X, Eggen BJ, Boddeke HW, Hambardzumyan D, Wolf SA, Kettenmann H. Glioma-associated microglia/macrophages display an expression profile different from M1 and M2 polarization and highly express Gpnmb and Spp1. PLoS ONE 2015;10, e0116644.

[51] Gabrusiewicz K, Rodriguez B, Wei J, Hashimoto Y, Healy LM, Maiti SN, Thomas G, Zhou S, Wang Q, Elakkad A, Liebelt BD, Yaghi NK, Ezhilarasan R, Huang N, Weinberg JS, Prabhu SS, Rao G, Sawaya R, Langford LA, Bruner JM, Fuller GN, Bar-Or A, Li W, Colen RR, Curran MA, Bhat KP, Antel JP, Cooper LJ, Sulman EP, Heimberger AB. Glioblastoma-infiltrated innate immune cells resemble M0 macrophage phenotype. JCI Insight 2016;1.

[52] Kessenbrock K, Plaks V, Werb Z. Matrix metalloproteinases: regulators of the tumor microenvironment. Cell 2010;141:52–67.

[53] Ye XZ, Xu SL, Xin YH, Yu SC, Ping YF, Chen L, Xiao HL, Wang B, Yi L, Wang QL, Jiang XF, Yang L, Zhang P, Qian C, Cui YH, Zhang X, Bian XW. Tumor-associated microglia/macrophages enhance the invasion of glioma stem-like cells via TGF-beta1 signaling pathway. J Immunol 2012;189:444–53.

[54] Plate KH, Breier G, Weich HA, Risau W. Vascular endothelial growth factor is a potential tumour angiogenesis factor in human gliomas in vivo. Nature 1992;359:845–8.

[55] Osterberg N, Ferrara N, Vacher J, Gaedicke S, Niedermann G, Weyerbrock A, Doostkam S, Schaefer HE, Plate KH, Machein MR. Decrease of VEGF-A in myeloid cells attenuates glioma progression and prolongs survival in an experimental glioma model. Neuro-Oncology 2016;18:939–49.

[56] Brandenburg S, Muller A, Turkowski K, Radev YT, Rot S, Schmidt C, Bungert AD, Acker G, Schorr A, Hippe A, Miller K, Heppner FL, Homey B, Vajkoczy P. Resident microglia rather than peripheral macrophages promote vascularization in brain tumors and are source of alternative pro-angiogenic factors. Acta Neuropathol 2016;131:365–78.

[57] Zhu C, Chrifi I, Mustafa D, van der Weiden M, Leenen PJM, Duncker DJ, Kros JM, Cheng C. CECR1-mediated cross talk between macrophages and vascular mural cells promotes neovascularization in malignant glioma. Oncogene 2017;36:5356–68.

[58] Hu F, Dzaye O, Hahn A, Yu Y, Scavetta RJ, Dittmar G, Kaczmarek AK, Dunning KR, Ricciardelli C, Rinnenthal JL, Heppner FL, Lehnardt S, Synowitz M, Wolf SA, Kettenmann H. Glioma-derived versican promotes tumor expansion via glioma-associated microglial/macrophages Toll-like receptor 2 signaling. Neuro-Oncology 2015;17:200–10.

[59] Coniglio SJ, Eugenin E, Dobrenis K, Stanley ER, West BL, Symons MH, Segall JE. Microglial stimulation of glioblastoma invasion involves epidermal growth factor receptor (EGFR) and colony stimulating factor 1 receptor (CSF-1R) signaling. Mol Med 2012;18:519–27.

[60] Zhang I, Alizadeh D, Liang J, Zhang L, Gao H, Song Y, Ren H, Ouyang M, Wu X, D'Apuzzo M, Badie B. Characterization of arginase expression in glioma-associated microglia and macrophages. PLoS ONE 2016;11, e0165118.

[61] Fonseca AC, Romao L, Amaral RF, Assad Kahn S, Lobo D, Martins S, Marcondes de Souza J, Moura-Neto V, Lima FR. Microglial stress inducible protein 1 promotes proliferation and migration in human glioblastoma cells. Neuroscience 2012;200:130–41.

[62] Matias D, Dubois LG, Pontes B, Rosario L, Ferrer VP, Balca-Silva J, Fonseca ACC, Macharia LW, Romao L, TCLS ES, Chimelli L, Filho PN, Lopes MC, Abreu JG, Lima FRS, Moura-Neto V. GBM-derived Wnt3a induces M2-like phenotype in microglial cells through Wnt/beta-catenin signaling. Mol Neurobiol 2019;56:1517–30.

[63] Nijaguna MB, Patil V, Urbach S, Shwetha SD, Sravani K, Hegde AS, Chandramouli BA, Arivazhagan A, Marin P, Santosh V, Somasundaram K. Glioblastoma-derived macrophage colony-stimulating factor (MCSF) induces microglial release of insulin-like growth factor-binding protein 1 (IGFBP1) to promote angiogenesis. J Biol Chem 2015;290:23401–15.

[64] Ding AS, Routkevitch D, Jackson C, Lim M. Targeting myeloid cells in combination treatments for glioma and other tumors. Front Immunol 2019;10:1715.

[65] Macauley MS, Crocker PR, Paulson JC. Siglec-mediated regulation of immune cell function in disease. Nat Rev Immunol 2014;14:653–66.

[66] Santegoets KCM, Gielen PR, Bull C, Schulte BM, Kers-Rebel ED, Kusters B, Bossman S, Ter Laan M, Wesseling P, Adema GJ. Expression profiling of immune inhibitory Siglecs and their ligands in patients with glioma. Cancer Immunol Immunother 2019;68:937–49.

[67] Schmid MC, Khan SQ, Kaneda MM, Pathria P, Shepard R, Louis TL, Anand S, Woo G, Leem C, Faridi MH, Geraghty T, Rajagopalan A, Gupta S, Ahmed M, Vazquez-Padron RI, Cheresh DA, Gupta V, Varner JA. Integrin CD11b activation drives anti-tumor innate immunity. Nat Commun 2018;9:5379.

CHAPTER

8

Indoleamine 2,3-dioxygenase 1 (IDO): A mediator of immunoresistance in adults with brain cancer treated with immunomodulatory therapy

Lijie Zhai[a], *Erik Ladomersky*[a], *Kristen L. Lauing*[a], *Lakshmi Bollu*[a], *April Bell*[a], *Miri Kim*[b], *Erik Rabin*[a], *and Derek A. Wainwright*[a,c,d,e]

[a]Department of Neurological Surgery, Northwestern University Feinberg School of Medicine, Chicago, IL, United States [b]Department of Neurological Surgery, Loyola University Chicago Medical Center, Maywood, IL, United States [c]Department of Medicine, Division of Hematology and Oncology, Northwestern University Feinberg School of Medicine, Chicago, IL, United States [d]Robert H. Lurie Comprehensive Cancer Center, Northwestern University, Chicago, IL, United States [e]Department of Microbiology-Immunology, Northwestern University Feinberg School of Medicine, Chicago, IL, United States

Abstract

Indoleamine 2,3-dioxygenase 1 (IDO; IDO1) is expressed in >90% of all adult-resected glioblastoma (GBM) and potently suppresses the anti-GBM immune response. IDO expression is increased by GBM-infiltrating T cells and therefore represents a key mechanism of resistance to immune-based therapies, since targeted approaches that aim to enhance T-cell infiltration into GBM will in turn result in a compensatory increase of IDO expression and suppression of the anti-GBM immune response. Importantly, the high intratumoral IDO expression is associated with a significant decrease of GBM patient survival and increased immunosuppressive regulatory T cells ($CD4^+CD25^+FoxP3^+$). Since IDO is canonically characterized as a rate-limiting

https://doi.org/10.1016/B978-0-12-819755-4.00006-0

enzyme that metabolizes the essential amino acid, tryptophan (Trp), into kynurenine (Kyn), the depletion of Trp and/or accumulation of Kyn has been the presumed mechanism of how IDO suppresses the anti-GBM immune response. However, our work has challenged this hypothesis, and here we provide a comprehensive update of IDO effects in adults with GBM.

Abbreviations

3-HAA	3-hydroxyanthranilate
3-HK	3-hydroxy-L-kynurenine
AhR	aryl hydrocarbon receptor
BMSCs	bone marrow-derived mesenchymal stromal cells
CA	cinnabarinic acid
CTLA-4	cytotoxic T-lymphocyte-associated protein-4
D1-MT	1-methyl-D-tryptophan
DCs	dendritic cells
GBM	glioblastoma
GCN2	general control nonderepressible 2
GITR	glucocorticoid-induced TNFR-related protein
ICB	immune checkpoint blockade
IDO	indoleamine 2,3-dioxygenase 1
IDO1	indoleamine 2,3-dioxygenase 1
IDOi	indoleamine 2,3-dioxygenase 1 inhibitor
IFN-γ	interferon-gamma
iTreg	inducible regulatory T cell
KA	kynurenic acid
Kyn	kynurenine
LPS	lipopolysaccharide
LTS	long-term survival
MDMs	monocyte-derived macrophages
mOS	median overall survival
mTOR	mammalian target of rapamycin
NAD^+	nicotinamide adenine dinucleotide
NFK	*N*-formylkynurenine
NF-κB	nuclear factor-kappa B
nTreg	natural regulatory T cell
PARP	poly-ADP-ribosyltransferase
PD-1	programmed cell death protein 1
PD-L1	programmed death ligand 1 (B7-H1)
PKC	protein kinase C
PTEN	phosphatase and tensin homolog
Quin	quinolinic acid
rGBM	recurrent glioblastoma
SASP	senescence-associated secretory phenotype
SHP	tyrosine phosphatase
TGF-β	transforming growth factor-beta
Th1	T helper 1
Th2	T helper 2
TLR9	toll-like receptor 9
TNF-α	tumor necrosis factor-alpha
Treg	regulatory T cell ($CD3^+CD4^+CD25^+FoxP3^+CD127^{low}$)
Trp	tryptophan
XA	xanthurenic acid

Conflict of interest

No potential conflicts of interest were disclosed.

Introduction

The immune system is composed of an immediate-acting innate arm comprised principally of granulocyte- and myeloid-lineage cells that respond to cues of inflammation and/or injury, in addition to an adaptive arm that's principally comprised of T and B cells capable of providing exquisite specificity and memory to previous immunological challenges. Under normal circumstances, the innate and adaptive arms are mutually dependent on one another for providing defense against infection, injury, and/or malignancy. T cells, which primarily mature following immunological challenge(s), include the $CD4^+$ and $CD8^+$ T-cell subsets and express a wide variety of cytokines that depend on the context of priming stimuli. The $CD4^+$ T-cell compartment also includes the highly immunosuppressive regulatory T cells (Tregs; $CD4^+CD25^+FoxP3^+CTLA\text{-}4^+GITR^+$) that mature naturally in the thymus (nTreg) or postthymically as a result of conversion from naïve $CD4^+Foxp3^-$ cells into Foxp3-expressing cells (iTreg) [1–3]. Solid cancers possess potently immunosuppressive mechanisms that evade antitumor immunity and often include Treg accumulation [4, 5], effector T-cell expression of the PD-1 receptor [6], and high PD-L1 levels in the tumor microenvironment [7, 8]. Therefore, ongoing efforts include the development of strategies that both reactivate an endogenous antitumor effector T-cell response while also simultaneously inhibiting immunosuppressive mechanisms.

Recent studies have demonstrated great promise at targeting immunosuppression in cancer including clinical trials aimed at inhibiting PD-1, PD-L1, and/or CTLA-4 in patients diagnosed with late-stage melanoma, nonsmall-cell lung cancer, and/or renal cell cancer [9–12] among a growing list of other aggressive cancer types. Follow-up studies have also shown that the benefit of combined PD-1/CTLA-4 inhibition is not restricted to those patients previously treated with systemic therapy [13]. Preclinical work using multiple tumor models in immunocompetent mice further confirms that these immune checkpoint-targeted therapies require effector T cells for antitumor activity, with several studies reporting a coincident neutralization of tumor-infiltrating Tregs [14–16]. These clinical studies, combined with extensive preclinical validation of combinatorial approaches, confirm that immunotherapy is a high-value strategy for treating patients with aggressive and immunosuppressive malignancies.

IDO, Trp metabolism, and its association with suppressing the anticancer immune response

Less than 1% of dietary Trp is used for protein synthesis under physiological conditions, while the remainder is degraded through decarboxylation, transamination, hydroxylation, or oxidation [17] leading to the generation of physiologically active compounds, including neuroactive tryptamine, neuroprotective melatonin, and/or immunomodulatory kynurenines. IDO catalyzes the rate-limiting cleavage of the Trp indole ring 2, 3-double bond and incorporates molecular oxygen. The product of this reaction is *N*-formylkynurenine (NFK) that is rapidly converted into Kyn by formamidase (Fig. 1). The latter catabolite is further transformed into downstream kynurenine-like intermediates, including 3-hydroxy-L-kynurenine (3-HK), 3-hydroxyanthranilate (3-HAA), and quinolinic acid (Quin), which also affect the immune response [19]. The ultimate product of the Kyn pathway metabolism is nicotinamide adenine dinucleotide (NAD^+), a critical enzyme cofactor involved in a variety of fundamental biological events, including acting as an electron transfer factor during glycolysis and oxidative phosphorylation. In addition, it is consumed by covalent modification of proteins and signaling enzymes. Monomeric IDO can act on a broad range of substrates that

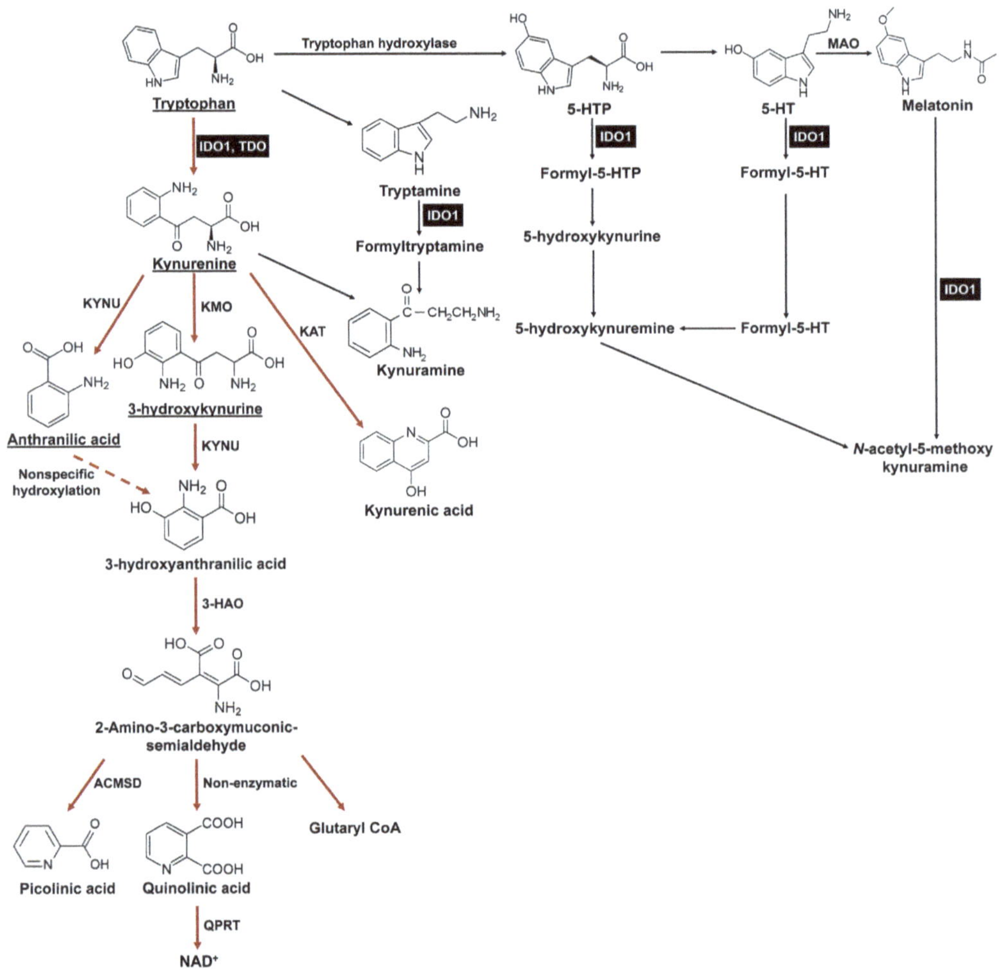

FIG. 1 The tryptophan (Trp)→kynurenine (Kyn) metabolic pathway [18]. The majority of dietary tryptophan (95%) is metabolized via the Trp➔Kyn pathway *(red arrows)*. A minor pathway that converts tryptamine into kynuramine is also included. Metabolites capable of crossing the blood-brain barrier (BBB) are *underlined*. IDO and TDO are highlighted in *black boxes*. *KMO*, kynurenine 3-monooxygenase; *KYNU*, kynureninase; *KAT*, kynurenine aminotransferase (I, II, III); *3-HAO*, 3-hydroxyanthranilate 3,4-dioxygenase; *ACMSD*, 2-amino-3-carboxymuconate semialdehyde carboxylase; *QPRT*, quinolinic-acid phosphoribosyl transferase.

include L-Trp, D-Trp, tryptamine, 5-hydroxytryptophan, and 5-hydroxytryptamine. The affinity of IDO for Trp and Trp-derivative substrates is highly variable such that binding to D-Trp is low as compared with L-Trp with a K_m of human IDO for the D-isomer that is 200-fold higher than that for the L-isomer [20, 21]. In contrast to IDO, the third Trp catabolic enzyme, IDO2, possesses a negligible level of intrinsic enzymic potential. The K_m of human IDO and IDO2 for L-Trp is $20.90 \pm 3.95\,\mu mol/L$ and $6809 \pm 917\,\mu mol/L$, respectively [22], supporting the hypothesis that IDO2 contributes a minimal role to overall Trp metabolism in humans.

Proinflammatory mediators tend to be the strongest agonists of IDO transcription that include the prototypical antitumor-associated T-cell effector cytokine, interferon-gamma (IFN-γ), as well as the toll-like receptor 9 (TLR9) and TLR4 agonists, CpG DNA, and lipopolysaccharide (LPS), respectively [23–26]. Tumor necrosis factor-alpha (TNF-α), interleukin-6 (IL-6), and IL-1β also synergize with one another to facilitate the upregulation of IDO expression. Other modulators include soluble glucocorticoid-induced TNFR-related protein (GITR), prostaglandin E2, the oncogene c-Kit, as well as the tumor suppressor, Bin1 [27]. The soluble fusion protein, cytotoxic T lymphocyte-associated protein 4-immunoglobulin (CTLA-4-Ig), induces IDO expression in dendritic cells (DCs) through the interaction of costimulatory receptors, CD80 (B7.1) and CD86 (B7.2) [28]. Previous work also confirmed that Wnt5α is an enhancer of IDO activity during β-catenin signaling in DCs [29].

Constitutive IDO expression can be found in healthy human subjects but is highly restricted to a select number of tissues and cells, including mature DCs in secondary lymphoid organs, epithelial cells inside the female genital tract, as well as endothelial cells of term placenta and lung parenchyma [30]. It should be noted that this superficial analysis does not capture the potential expression changes that occur across age groups as we previously found in the normal human brain parenchyma during advanced age [31]. While some human tumor cell types constitutively express IDO, its presence is normally absent in human and mouse glioblastoma cells but is rapidly induced upon treatment with human and mouse IFN-γ, respectively [32–36].

The premise for the tryptophan metabolism-induced immune suppression hypothesis is based on the near-absolute depletion of Trp at or below $<1\,\mu M$, which facilitates the accumulation of uncharged tRNAs that then activate the general control nonderepressible 2 (GCN2) kinase pathway and the dysfunction of T cells [37]. In vitro studies support the hypothesis that Trp depletion inhibits the master metabolic regulator mammalian target of rapamycin (mTOR) and protein kinase C (PKC-Θ) in cancer cells, which consequently enhances autophagy and Treg development, respectively [38]. Trp degradation may also suppress immune cell activities through the formation of Kyn and downstream derivative metabolites. Additionally, cotreatment with transforming growth factor-beta (TGF-β) and Kyn facilitates the induction of FoxP3 in naïve $CD4^+$ T cells by activating the aryl hydrocarbon receptor (AhR) [39], a ligand-activated transcription factor that exerts potent effects on immune cells [40] and is involved in the differentiation of iTregs [41, 42]. The downstream pathway Kyn metabolites, including kynurenic acid (KA) [43], xanthurenic acid (XA) [44], and cinnabarinic acid (CA) [45], interact with AhR and may also play a role in modulating the immune response. Trp catabolites have also been demonstrated to induce $CD4^+$ T-cell apoptosis. Terness et al. found that Kyn, 3-HK, and 3-HAA suppress T-cell proliferation coincident with the induction of apoptosis [46]. This finding was independently confirmed by Fallarino et al. [47] that demonstrated Kyns induce the selective apoptosis of murine thymocytes and

Th1, but not Th2, cells in vitro. The immunoregulatory role of Kyns on different lymphocyte subsets may be important for maintaining peripheral lymphocyte homeostasis and for minimizing the accumulation of autoreactive and/or inflammatory lymphocytes. It's important to note that the majority of Trp- and Kyn-mediated immunoregulatory effects have only been observed in vitro.

Since Trp is an essential amino acid and not readily generated by mammalian cells, dietary consumption is critical for host survival and in vivo replenishment. It is therefore important to consider the Trp concentrations required to inhibit T-cell proliferation in vitro, which are below 0.5–1 μM [37]. Whether cell culture conditions can represent physiological conditions in vivo is controversial since human plasma Trp levels normally range between 50 and 100 μM. A number of conflicting results on this topic have recently arisen and were highlighted by Frumento et al. during their failure to observe an inhibition of T lymphocyte proliferation when the in vitro cell culture media were absent for Trp [48]. Although it is arguable that high expression of tumor cell IDO might lead to extremely low Trp levels within the tumor microenvironment, it's challenging to validate this hypothesis in vivo.

Another hypothetical explanation regarding Trp depletion relies on the alteration of the Trp transporter activity. This theory suggests that even while there is no dramatic decrease of Trp in the peripheral blood circulation, alteration of the Trp transporter activity may cause intracellular Trp depletion, which triggers downstream immunosuppressive signals. In addition to the well-known nonspecific low-affinity Trp transporter (system L) [49], two Trp-specific transporters with high affinity have been reported to be expressed on monocyte-derived macrophages and tumor cell lines [50, 51]. Unfortunately, neither of these novel Trp transporters have been successfully cloned to date. Comprehensive functional analyses of the Trp transporters expressed by different immune cells and/or tumor cells in vitro and in vivo remain largely unexplored.

Despite the elegant in vitro work by Mezrich et al. reporting that the 50 μM Kyn treatment activates AhR in naïve T cells and subsequently induces FoxP3 expression [39], human serological Kyn levels are normally within the range of 2–3 μM in both healthy individuals and patients with inflammatory conditions [52–56]. Theoretically, local tissue Kyn levels might transiently increase when IDO is highly expressed [57], but more precise in vivo detection and quantification techniques remain unavailable. Conflicting with the report by Mezrich et al., Nguyen et al. reported that the addition of Kyn in a DC-T-cell coculture system led to the inhibition of naive T-cell differentiation into $FoxP3^{+}$ Tregs [58]. This latter observation supports the hypothesis that the Kyn-AhR mechanism is not responsible for the direct generation of Tregs in solid tumors, especially since the infiltrating components of tumors tend to primarily consist of thymus-derived natural Tregs [59, 60] that do not express Ahr [61]. It's also notable that other studies indicate the potential role of Kyn on Treg activation as evidenced by the Munn and Mellor group who showed IDO-expressing plasmacytoid DCs in tumor-draining lymph nodes preferentially activate preexisting Tregs through the PD-1-PTEN pathway which is inhibited by the IDO pathway inhibitor, 1-methyl-D-tryptophan (D1-MT) [62, 63]. In addition to the sometimes conflicting outcomes discussed above, results from studies suggesting that Kyns directly suppress the actions of effector T cells have also been revisited due to the high doses previously evaluated, which ranged from 100 to 1000 μM [46, 47]. The panoply of different observations across various modeling systems highlights the complexity of IDO immunobiology that may be conveyed as artefactual effects during in vitro study and thereby supports the re-evaluation of previous interpretations with the use of new physiologically relevant in vivo modeling systems.

Pleiotropic IDO effects reflect multiple functions, diverse cellular origins, and varying kinetics

In addition to the highly restricted tissues that constitutively express IDO, the induction of its expression occurs in a large variety of cell types under pathological conditions. This includes IDO expression changes during progressive aging [31], as well as in the setting of malignancy whereby levels change in response to T-cell infiltration [64]. The complexity of IDO functionality has been highlighted by immunohistochemical IDO detection across 15 different human cancer types that found diverse protein localization among myeloid cells, endothelial cells, tumor cells, or a combination of those origins [30]. The pattern of intratumoral IDO expression raises critical questions, including: (i) Does IDO perform the same role among different cell types; (ii) how do these cells coordinately contribute to tumor growth and/or suppression of anticancer immunity; (iii) what are the kinetics of endogenous IDO expression and its relationship to enzyme activity?

A peculiar trait of tumor cells expressing high IDO expression is the resulting effect on an overall slower rate of growth as compared with cells with lower intrinsic IDO levels [65]. A derivative and competing hypothesis arising from this observation suggests that IDO functions as a tumor suppressor. This was corroborated with findings of higher intratumoral IDO expression positively associated with longer survival in renal cell carcinoma, hepatocellular carcinoma, and melanoma patients [18, 66–68]. It should also be considered, however, that higher levels of T-cell infiltration are often associated with higher intratumoral IDO levels, and higher intratumoral T-cell levels are associated with an increased chance of survival among human subjects with select malignancies such as melanoma [18, 69]. Therefore, the higher intratumoral IDO levels are more likely to be an associative effect and not a direct effect on tumor growth itself. In contrast to the alleged effects of slower growth, higher IDO expression was associated with an improved motility of lung cancer cells that enhanced their metastatic formation in the brain, liver, and bone [70], whereas IDO deficiency decreased metastatic burden and improved the survival of subjects with breast carcinoma-derived pulmonary metastases [71, 72]. Intratumoral IDO expression also correlated with the frequency of liver metastases in colorectal cancer [73], distant metastases in hepatocellular cancer [68], and nodal metastases in endometrial carcinoma [74]. Among human cancer cells, IDO activity has been implicated in improving DNA repair and mediating the resistance to treatments such as the poly-ADP-ribosyltransferase (PARP) inhibitor, olaparib, γ-radiation, and the chemotherapeutic agent, cisplatin, through the generation of NAD^+ [75]. It was reported that impaired Trp metabolism resulted in the inhibition of de novo NAD^+ synthesis, which led to hepatic tumorigenesis through DNA damage [76], further supporting the previous finding that NAD^+ serves as the only endogenous substrate of PARPs to facilitate the removal of oxidative DNA damage. It's important to note, however, that IDO is not normally expressed in the liver.

While IDO expression by tumor and immune cells tends to receive the most attention, IDO is also expressed by endothelial cells [30]. In healthy tissues, endothelial cell IDO is present in a large proportion of placental and pulmonary blood vessels, as well as by a minority of blood vessels in select other organs. Endothelial IDO appears to modulate vascular tone through several different signaling pathways, including Kyn-mediated activation of soluble guanylate cyclase, adenylate cyclase, and voltage-dependent K^+ channels [77, 78]. The Kyn metabolite, xanthurenic acid, possesses greater potency as compared with L-Kyn in causing blood vessel relaxation that is dependent on nitric oxide [79]. Stanley et al. found that in the presence of H_2O_2, IDO catalyzes Trp oxidation using singlet molecular oxygen ($^1O_2$) that leads to the vasodilation product *cis*-WOOH [80]. It's important to note that these experiments were conducted while utilizing high concentrations of the Trp catabolites between 300 μM and 1.5 mM in vitro. It's therefore unclear as to whether the vasodilating compounds attain a similar local level inside a solid tumor in vivo. Previous work supports the vaso-relaxing effects of endothelial IDO including a model of sepsis-induced hypotension [81]. Intrinsic to the IDO expressed by endothelial cells, there may also be a role for an effect on modulating tumor neovascularization [82], though the exact molecular mechanism explaining this has yet to be elucidated [83].

Data suggesting that IDO possesses nonmetabolic functions independent from those associated with its enzyme activity began emerging ~10 years ago. Pallotta et al. were the first to report that the addition of TGF-β to cultures of mouse plasmacytoid DCs (pDCs) in vitro leads to the phosphorylation of IDO immunoreceptor tyrosine-based inhibitory motifs (ITIMs), which subsequently recruit and activate the tyrosine phosphatases, SHP-1 and SHP-2, as well as inositol polyphosphatase (SHIP) [84]. Activation of the pDC TGF-β-IDO-SHP axis enabled an autocrine loop through the induction of noncanonical NF-κB signaling, which further enhanced intra-pDC IDO expression levels. Treatment of the pDCs with 1-methyl-D,L-tryptophan confirmed that the immunosuppressive IDO signaling mechanism was faithfully independent of its association with mediating Trp metabolism. However, it's notable that the 4 μM dose of 1-MT used in this study was lower than the EC_{50} of 1-MT [85]. Our group found similar observations of IDO mediating both enzymatic and nonenzymatic effects that depend on the context. Using a syngeneic brain tumor model, we demonstrated that the shRNA knockdown of GBM cell IDO leads to the suppression of intratumoral Treg accumulation and is associated with a significant improvement in long-term animal subject survival [86] independent of GBM cell IDO metabolism [36]. We also showed that the forced expression of GBM cell IDO cDNA (IDO-O/E) enhances Treg recruitment even when animal subjects are treated with a potent blood-brain barrier-penetrating pharmacologic IDO enzyme inhibitor [87]. We further confirmed that while the IDO enzyme inhibitor significantly decreased intratumoral Kyn levels in GBM IDO-O/E in vivo, the reduction was not associated with decreased intratumoral Treg accumulation. Taken together, these observations support the hypothesis that new IDO-targeting approaches aimed at simultaneously reversing enzymatic and nonenzymatic activities will enhance the effectiveness of future cancer immunotherapy efforts.

IDO and its relationship with host age

Considering the immunosuppressive role of IDO, as well as the immunological atrophy during progressive aging, it's not surprising that several studies have investigated the relationship between the Trp→Kyn pathway and aging-dependent disease. The serological

Kyn/Trp ratio is often used as an indicator of IDO enzymatic activity and progressively increases during normal aging in humans [88, 89]. This increase has been associated with enhanced frailty in human subjects >65 years of age and predicts an increased mortality rate of individuals in their 1990s [90]. Additionally, metaanalysis of age-related gene expression changes in the peripheral blood of adult individuals identified the enzyme kynureninase as one of the most differentially expressed genes [91]. In a large population screening analysis that included 7074 human subjects focused on studying bone mineralization, Apalset et al. reported that the serological Kyn/Trp ratio negatively correlated with bone mineral density in the 71- to 74-year-old age group as compared with younger individuals that were 46–49 years of age [92]. More recently, Ocampo et al. summarized the results from previous animal studies of brain diseases and the interaction with Kyns during aging that indicate a strong correlation between Kyn pathway metabolites across different rodent ages [93]. Similarly, our group previously demonstrated that there is an ~400-fold increase in IDO mRNA expression in the normal naïve brain of 72- to 74-week-old C57BL/6 WT mice as compared with 6- to 8-week-old subjects [94]. Our most recent study evaluating the interactions between normal human aging and its relationship with brain cancer incidence and mortality also found a significant increase in IDO mRNA expression in the normal human brain of individuals aged 60–69 years as compared with younger human subjects [31]. The increased human brain IDO expression was associated with a maximal Treg/CD8$^+$ cytolytic T-cell ratio in the peripheral blood of the 60- to 69-year-old age group, confirming a simultaneous increase of systemic and local immune suppression in the brain [31]. To address these observations, we are currently in the process of determining as to whether the advanced age-dependent increase of IDO expression in the central nervous system directly increases the peripheral Treg/CD8$^+$ cytolytic T-cell ratio and how this relationship affects the incidence and mortality rate, as well as responsiveness to immunotherapy of GBM.

Understanding how aging-dependent molecular mechanisms affect IDO enzymatic and nonenzymatic activity is in its infancy. Refaey et al. reported that 22-month-old mice have significantly higher serum *N*-formylkynurenine as compared with younger counterparts at 4 and 13 months of age, suggesting a potential role for Kyn pathway metabolism in aging-dependent bone loss [95]. Strikingly, either the in vitro addition of Kyn to bone marrow-derived mesenchymal stromal cells (BMSCs), dietary supplementation of Kyn, or a direct administration of Kyn into the peritoneum suppressed the formation of new bone, possibly by decreasing osteoblast formation, failed recruitment capability, and/or loss of effector functions. It should be noted, however, that age-matched IDOKO mice were not used in the study and it's therefore inconclusive as to whether it was the primary source of aging-dependent changes in Kyn levels. Independently, Minhas et al. demonstrated that the catabolic enzyme converting quinolinic acid into NAD$^+$, quinolate phosphoribosyltransferase, significantly declines in aged human monocyte-derived macrophages (MDMs) from individuals ≥65 years of age as compared with MDMs isolated from individuals ≤35 years of age [96]. Loss of quinolate phosphoribosyltransferase subsequently resulted in decreased de novo NAD$^+$ synthesis and was associated with an enhanced proinflammatory status and lower phagocytic ability of MDMs. It's worth noting that in addition to the enzymes and metabolites of the Kyn pathway, changes in AhR activity and/or expression have also been implicated during aging [97]. However, an investigation of the direct relationship between IDO, Kyn, and AhR has not been established in a specific pathophysiological setting to date.

Adult glioblastoma is an age-related disease with a median age of diagnosis at 65 years [98]. To explore the mechanistic underpinnings between advanced age and its effects on the GBM immune response, we previously compared the survival rate of young 6- to 8-week- and adult 72- to 74-week-old mice intracranially engrafted with the syngeneic GL261 glioma cell line. The young mice survived slightly longer than the older subjects with a median overall survival (mOS) of 27.5 days and 21.5 days, respectively ($P = .0292$) [94]. Intriguingly, advanced age was associated with a large increase of IDO expression in the contralateral brain without tumor and as compared with young mice, whereas there was no such difference within the tumor itself. These data suggest that the majority of brain parenchymal cells expressing increased IDO during advanced age do not migrate into the GBM. Intriguingly, serological and intra-GBM Trp and Kyn levels showed no differences between young and adult subjects despite the increased IDO expression in the older brain [94]. We also studied the effects of advanced age on immunotherapeutic efficacy in adult mice at 72 weeks of age, which is closer to the analogous time frame of a human GBM patient diagnosis as compared with the standard 6- to 12-week-old mouse. When the young and adult mice underwent treatment [87] with whole brain radiotherapy, anti-PD-1 mAb and pharmacological IDO enzyme inhibitor, older mice showed higher IDO mRNA in the nontumor brain tissues as compared with younger subjects. Functionally, the triple combination treatment resulted in a median survival of 31.5 days in the older subjects, which was significantly decreased as compared with the younger 8-week-old mice with a median survival of 40.5 days ($P < .001$), suggesting that the increased brain IDO expression during advanced age has a directly negative effect on immunotherapeutic efficacy in subjects with GBM.

Mouse models for studying IDO

Among preclinical studies, transgenic mouse models have demonstrated their substantial importance to understanding the function, effects, and mechanism of IDO in a physiologically relevant environment. The first IDO-knockout (IDOKO) mouse strain originated from the Munn and Mellor group whereby exons 3–5 of IDO were replaced with a β-gal and neomycin cassette [99, 100]. Surprisingly, the homozygous IDOKO mice were viable, fertile, and possess a phenotypically normal immune system without any obvious graft-versus-host-like symptoms, suggesting that IDO-mediated immune suppression is dispensable for the normal maintenance of central and peripheral tolerance to self-antigens. Though IDO expression is not required for immune tolerance under homeostatic conditions, it plays a critical role during the acquisition of tumor immune tolerance as indicated by studies demonstrating that IDOKO mice have increased tumor infiltration of effector T lymphocytes, decreased immunosuppressive immune cells, as well as increased survival [101–104], implicating nonredundant functions between IDO and TDO in the various cancer settings.

Because of their normal growth and immune system development, constitutional IDOKO mice are widely used preclinically. However, this transgenic model does not reveal the role of IDO in different tissues or cell types. Recently, Bishnupuri et al. generated an intestinal epithelium-specific IDO-knockout mouse strain by breeding *Ido1*$^{tm1c(EUCOMM)Wtsi}$ mice with the villin promoter *Cre*-expressing mouse [104]. The mouse colon cancer model utilizing this conditional IDO-knockout mouse strain discovered that neoplastic colon epithelium cell IDO is a critical constituent of colon tumorigenesis and that Kyn metabolites rapidly activate PI3K-Akt signaling in the neoplastic epithelium to promote cellular proliferation and resistance to apoptosis. In addition to the constitutional and conditional IDO-knockout mice that have previously been reported, our group has created new constructs that will allow for addressing novel questions including an IDO reporter mouse that expresses IDO-GFP (Fig. 2A), an IDO-knock-in mouse that allows for crossing a tissue-specific Cre-expressing mouse with our recently created Rosa26Sor-loxp-IDO-FLAG-GFP (Fig. 2B), and an IDO

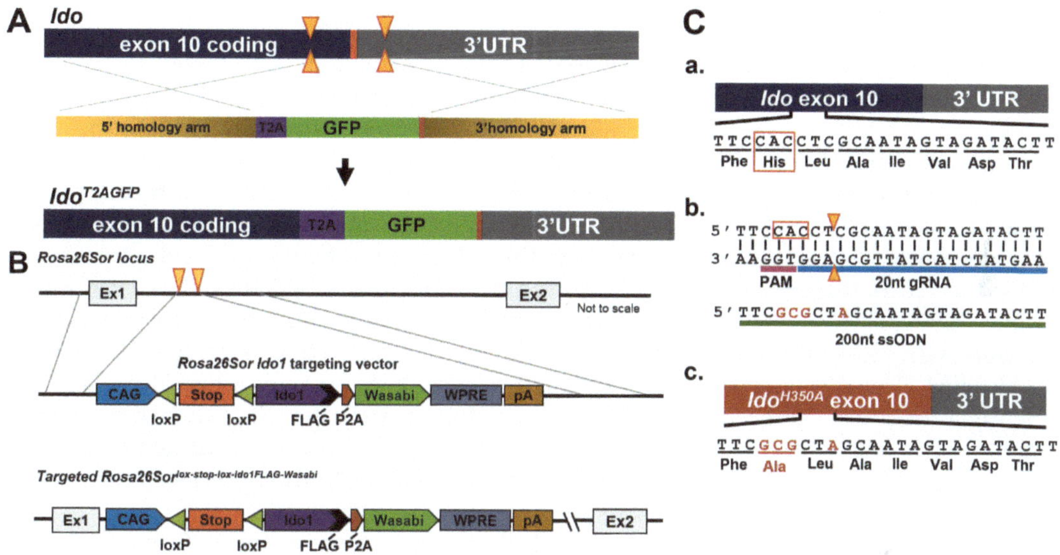

FIG. 2 Schema utilized to generate IDO-targeted transgenic mouse strains [105]. (A) Generation of the *Ido1*^{T2AGFP} mouse line. CRISPR/Cas9 technology was employed in murine embryonic stem (ES) cells. Two separate guide RNAs directed Cas9 nuclease cleavage of the *Ido1* locus, creating double strand DNA breaks *(yellow arrow)* before and after the stop codon *(red bar)*. The CRISPR ES cell reaction included a plasmid repair template, which encoded the T2A and GFP elements flanked by 5′ and 3′ homology arms of 700 nt *(tan bars)*. The T2A and GFP elements were introduced into the endogenous *Ido1* locus before the stop codon via homology-directed repair (HDR). (B) Introduction of a *Cre*-dependent *Ido1* expression construct into the *Rosa26Sor* locus by CRISPR/Cas9 technology. Two separate guide RNAs directed Cas9 nuclease cleavage at the *Rosa26Sor* locus, creating double-strand DNA breaks *(yellow arrows)*. A plasmid repair temple (*Rosa26Sor Ido1* targeting vector) was included in the reaction and introduced via HDR. The edited locus encodes the CAG promoter-driven, *Cre*-dependent conditional expression of a bi-cistronic *Ido1* construct. The *Ido1* transcript includes a C-terminal FLAG-tag, followed by a P2A element and expression of the Wasabi fluorophore. It also included a WPRE element and the bovine growth hormone polyadenylation (pA) signal. The construct is flanked by 800-bp 5′ and 750-bp 3′ homology arms *(gray lines)*. (C) Generation of the *Ido1*^{H350A} mouse line. (a) Schematic showing the last coding exon of the *Ido1* gene, exon 10. The highlighted region encodes a histidine residue at position 350 *(red box)* that is essential for mouse IDO enzyme activity; (b) CRISPR/Cas9 technology generated a precise double-strand DNA break in the *Ido1* exon 10 locus in fertilized mouse embryos *(yellow triangle)*. A 200-nt single-stranded oligodeoxynucleotide repair template was included in the reaction *(green bar)*. The ssODN introduced point mutations *(red letters)* into the endogenous *Ido1* locus via HDR. These point mutations altered the endogenous coding sequence from CAC → GCG, which changed the amino acid reside from histidine to alanine, H350A *(red letters)*. An additional silent mutation (C → A) was included to add a restriction site for genotype analysis *(red letter)*; (c) Schematic showing the mutated *Ido1* exon 10 locus encoding H350A in *red*.

enzyme-nullified mouse whereby the histidine of the 350th amino acid is substituted to an alanine (Fig. 2C). We are currently characterizing each of these models and will publish data supporting their phenotypes in the near future.

The 10,000 ft view of IDO's role in adults with glioblastoma

Immune checkpoint blockade treatment with anti-PD-1 mAb has demonstrated promise in several large randomized phase III clinical trials for patients diagnosed with end-stage melanoma [13], nonsmall-cell lung [106] or renal cell [107] cancer, in addition to many others. Unfortunately, neither newly diagnosed [108] nor recurrent [109] GBM patients have benefitted from ICB treatment among phase III trials to date. Coincident with this observation, it's now well recognized that T-cell infiltration increases the expression of IDO in human GBM. Previous work by Dr. Donald O'Rourke and colleagues found that a single dose of systemically infused EGFRvIII-specific CAR T cells into a patient with $EGFRvIII^+$ GBM was associated with a postinfusion increase of $CD3\varepsilon^+CD8^+$ cytolytic T cells and immunosuppressive IDO expression inside the tumor [110]. During a similar time frame, our group published a complementary observation showing that the infiltration of human T cells into human GBM led to the increased expression of intratumoral IDO levels. Specifically, we utilized humanized NSG-SGM3-BLT mice with human immune systems that we then intracranially engrafted with either human U87 GBM, human patient-derived xenograft 12 (PDX12), or PDX43 GBM. Those mouse models were then treated with or without antihuman T cell-depleting antibodies. Our study demonstrated for the first time that (i) humanized mice with human GBM are a relevant model system for studying the intratumoral gene expression changes caused by infiltrating human T cells and (ii) GBM-infiltrating human T cells directly increase immunosuppressive IDO levels in the tumor parenchyma.

We previously demonstrated that higher IDO expression was associated with decreased glioma patient survival [15]. When we stratified glioma subtypes, high intratumoral IDO expression was specifically associated with a poor prognosis in patients with GBM [111]. Intratumoral IDO mRNA levels progressively increased with glioma grade, and the highest IDO expression levels were found in GBM [64] coincident with a higher IDO intratumoral expression level being associated with a faster rate to GBM recurrence in patients [112]. Coincident with the observations in human GBM, we also determined that GL261 cell-expressed IDO decreases the survival of syngeneic C57BL/6 mice [86]. The GBM cell IDO inhibited the ability of anti-GBM T cells to eradicate tumor cells. This conclusion was based on the finding that mice deficient for GBM cell IDO, but also deficient for T cells, lose all survival benefits related to tumor cell IDO deficiency. These were the first in vivo data to indicate that glioma cell IDO suppresses the anti-GBM T-cell response and promotes uncontrolled tumor outgrowth.

GBM cell IDO increases the number of immunosuppressive regulatory T cells (Tregs; $CD4^+CD25^+FoxP3^+$) inside the brain tumor, which in turn suppresses the anti-GBM immune response. Tregs accumulate in a large subset of newly diagnosed patient-resected GBM [113] and are present in nearly 100% of recurrent patient-resected GBM. Treg levels increase in parallel with histopathologic glioma grade—with grade IV glioma (GBM) possessing the highest levels [5]. Our group previously showed that GBM cell IDO increases intratumoral Treg levels [86] and that they enhance tumor outgrowth since the depletion of Tregs extends the survival of syngeneic C57BL/6 mice with intracranial GL261 [59, 114]. Similarly, Treg neutralization also enhances GBM patient vaccine efficacy [115]. Observations such as these have helped us to generate new hypotheses for how and why Tregs accumulate in GBM and suppress the antibrain tumor immune response (Fig. 3) [116]. Since GBM cell IDO is a central mediator of intratumoral Treg accumulation, it was logical to validate the presumed mechanism in vivo [117–119].

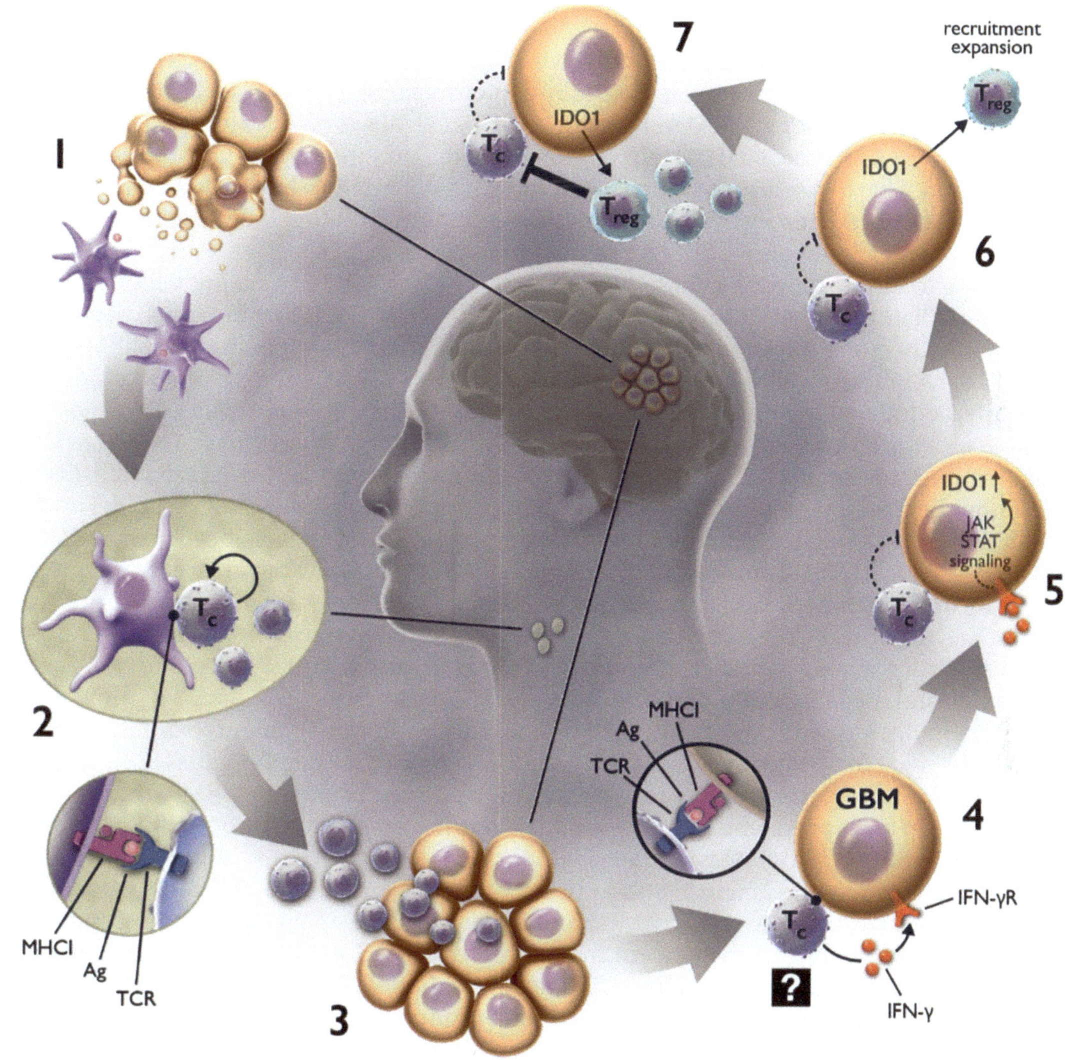

FIG. 3 A model explaining how cytolytic T cells induce IDO1 expression and immune suppression in glioblastoma (GBM) [64]. (1) GBM-derived antigens are taken up by tumor-infiltrating antigen-presenting cells (APCs), such as dendritic cells, myeloid macrophages. (2) APCs migrate to tumor-draining lymph nodes and activate cognate naïve cytolytic T cells (Tc) through MHC I-peptide:T-cell receptor interactions. (3) Activated Tc cells migrate to GBM. (4) Activated Tc cells secrete the effector antitumor cytokine, interferon gamma (IFN-γ). (5) IFN-γ binds to IFN-γ receptor (IFN-γR), which induces the expression of IDO1 expression via JAK-STAT signaling. (6) GBM cell IDO1 leads to regulatory T cells (Treg) accumulation and (7) suppresses the antitumor effector functions of Tc cells, which promotes GBM outgrowth.

Since the canonical function of IDO is to catalyze the rate-limiting step of tryptophan conversion into kynurenine, and because we previously showed that shRNA knockdown of GBM cell IDO decreases intratumoral Treg levels, we next hypothesized that a potent pharmacologic IDO enzyme inhibitor will inhibit Treg accumulation in GBM. We began our analysis by first comparing untreated syngeneic C57BL/6 mice with unmodified GL261 and GL261-expressing murine IDO cDNA. The murine IDO cDNA-expressing brain tumors had higher IDO mRNA levels, higher IDO enzyme activity based on the high Kyn/Trp ratio, and an increased number of intratumoral Tregs [87]. Next, we compared mice with intracranial GL261 IDO-OE treated with or without a pharmacologic IDO enzyme inhibitor. As expected, the pharmacologic IDO enzyme inhibitor had no effect on intratumoral IDO mRNA expression, and it decreased IDO enzyme activity based on the significantly decreased Kyn/Trp ratio. Unexpectedly, treatment with the IDO enzyme inhibitor had no effect on decreasing intratumoral Treg levels. Ultimately, these data confirm that GBM cell IDO increases intratumoral Tregs independent of tryptophan metabolism.

Wild-type isocitrate dehydrogenase (wtIDH) GBM reflects 90% of all adult GBM diagnoses and has a median age of onset at ~68–70 years old. It arises without any previous diagnosis of cancer and is therefore not the result of a lower-grade glioma transitioning into GBM. While the median overall survival (mOS) of patients with GBM following surgery, radiotherapy, chemotherapy, and tumor-treating fields is 18–21 months according to a recent phase III clinical trial [120], the mOS is less than 6 months for those GBM patients that are ≥65 years of age [121], highlighting an underappreciated aspect of this central nervous system age-related disease and raising the question of why GBM has a dramatically poorer prognosis among older adults? Advanced age is associated with a higher mortality rate [31], as well as decreased mOS and long-term survival (LTS) of subjects with GBM [122]. Although GBM can arise at any age, there is a significant enrichment of diagnoses among subjects ≥65 years of age. Coincident among this enriched population is a decreased mOS and fewer long-term survivors that are consistent with the hypothesis that advanced age increases factors in the brain that contribute to the onset, progression, and/or failure of immune system control of tumor burden.

Similar to older human subjects with GBM, older mice engrafted with intracranial malignant glioma have fewer long-term survivors and are more sensitive to lymphopenia [122]. To determine whether the negative effects of advanced age observed in human GBM patients can be recapitulated in an experimental model, we analyzed the mOS and LTS of C57BL/6 mice intracranially engrafted with 500 syngeneic CT-2A glioma cells at 6, 23, 47, and 75 weeks of age that were administered IgG control antibodies or anti-CD4 and anti-CD8 T cell-depleting antibodies. There were different levels of mOS and LTS depending on the age of subject engrafted with glioma cells such that there were higher levels of LTS among middle-aged subjects with T cells, whereas older adults had no LTS and their mOS level was decreased by the absence of T cells. These data collectively suggest the three primary conclusions: (i) T cells provide different levels of protection from glioma that depend on the subject's age; (ii) older adult T cells are critical for preventing a catastrophic decrease of mOS from glioma; and (iii) older adults with glioma have poorer LTS outcomes as compared with younger counterparts.

Older adult human patients with GBM have a decreased mOS after treatment with immune checkpoint blockade (ICB) [122]. Because we found a lower mOS among older human adults with GBM, we next questioned whether this effect extended to GBM patients treated with ICB such as anti-PD-1, anti-PD-L1, or anti-CTLA-4 mAbs. Information was extracted from medical records, including age, sex, MGMT promoter methylation status, and IDH status for patients with recurrent GBM (rGBM) that were administered ICB in the adjuvant setting. Ninety-two patients were dichotomized into groups defined as <65 years of age ($n = 65$) and ≥65 years of age ($n = 27$). rGBM patients <65 years of age treated with ICB had a higher mOS of 278 days as compared with patients ≥65 years of age that had a mOS of only 154 days ($P = .022$). Similar to human subjects with GBM, older adult mice with intracranial glioma have decreased survival as compared with younger counterparts following treatment with ICB. The negative association between advanced age and decreased mOS in human GBM patients, combined with the negative association between advanced age and decreased mOS in human rGBM patients treated with adjuvant ICB, prompted us to generate a table of major analogous age groups between wild-type C57BL/6 mice and humans. Based on our previous work determining the promising survival benefit of anti-CTLA-4 mAb/anti-PD-L1 mAb ICB treatment in young syngeneic mice [15], we next questioned how advanced age would affect the preclinical survival benefit. Similar to the clinical setting, older adult mice with glioma had a significantly decreased mOS of 25.5 days and 27% LTS as compared with younger subjects that had an undefined mOS and 67% LTS. The overlapping negative outcomes of advanced age in mouse and human subjects with malignant glioma suggest that factors exist in older adults, but not in younger subjects, that decrease the survival benefit of therapy.

Advanced age decreases therapeutic efficacy through a nontumor, nonhematopoietic cell-dependent mechanism [122]. To explore the basis for how advanced age decreases therapeutic efficacy in subjects with glioma, we next investigated bone marrow chimeric mice with different ages of host nontumor stroma and hematopoietic cells. Given our previous work validating the combination of radiation (RT), anti-PD-1 mAb, and an indoleamine 2,3-dioxygenase 1 enzyme inhibitor (IDOi) to improve the survival of young mice with well-established glioma [87], which serves as the basis for the Phase I clinical trial in newly diagnosed GBM patients [NCT04047706] led by Dr. Rimas Lukas, MD, we next evaluated this clinically relevant therapeutic strategy while assessing the major potential sources of immune suppression during advanced age. When comparing either the young WT mouse group or the old WT mouse group with glioma, no differences in mOS were found. In contrast, there was a significant increase in both the mOS and the LTS of young mice with glioma that were reconstituted with either young or older adult bone marrow, as compared with older adult mice that were reconstituted with either young or older adult bone marrow ($P < .0001$). These data collectively suggest that advanced age suppresses therapeutic efficacy against glioma through a nonhematopoietic, nontumor stromal-cell dependent mechanism. Coincident with this observation, immunosuppressive IDO increases in the brain during advanced age. We and others have demonstrated substantial immunological changes in the brain during advanced age. Notably, these changes occur independent of tumor burden and include increased human and murine [94] IDO expression.

IDO decreases therapeutic efficacy in older adults with glioma and its effects are not reversed by treatment with a metabolic inhibitor [122]. Our finding of higher IDO levels in the older adult brain combined with the survival-decreasing effects of IDO in older adults prompted us to next evaluate the effects of advanced age-increased IDO in a therapeutic setting. Young and old WT and IDOKO mice with intracranial glioma were treated with RT, anti-PD-1 mAb, and IDOi at 14 days after glioma cell engraftment. In support of our previous findings [87], there was a significant decrease in mOS and LTS in older adult WT mice as compared with younger counterparts ($P < .0001$). Also in line with our previous report, there was no difference of mOS or LTS in young WT and IDOKO mice with glioma treated with the

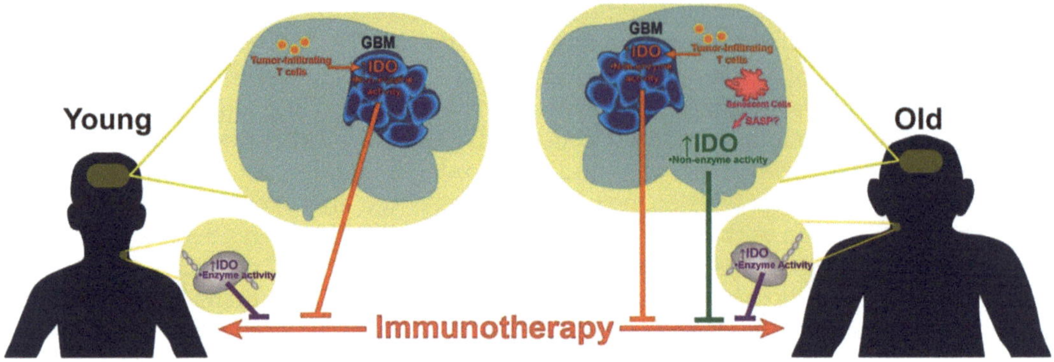

FIG. 4 A working hypothesis for how IDO suppresses the antiglioma immune response and/or responsiveness to glioma immunotherapy [122]. Greater than 90% of human patient-resected glioblastoma (GBM) is IDO positive and for which, intratumoral IDO expression levels increase with higher tumor-infiltrating T-cell levels but are unchanged with aging. Glioma cell IDO increases intratumoral immunosuppressive regulatory T cells (Tregs; $CD4^+FoxP3^+$) [86] despite the treatment with a pharmacological IDO enzyme inhibitor and the absence of IDO metabolism by brain tumor cells [36, 123]. In contrast, IDO metabolism is detectable in the brain tumor-draining cervical lymph nodes (cLN) [36]. Advanced aging is associated with increased IDO expression in the brain [94] and IDO decreases immunotherapeutic efficacy against glioma during advanced aging regardless of IDO enzyme inhibitor treatment. Given their association with mediating inflammation, senescent cells with the senescence-associated secretory phenotype (SASP) indirectly upregulate immunosuppressive IDO expression in the brain during advanced aging due to the high sensitivity for IDO expression by proinflammatory molecule expression [119].

triple combination therapy. In a marked contrast, older adult IDOKO mice treated with the triple combination showed an impressive improvement in both mOS and LTS as compared with the older adult WT counterparts ($P < .01$). The data collectively suggest that advanced age increases IDO expression in the brain and suppresses therapeutic efficacy independent of IDO enzyme activity.

Here, we present a working hypothesis for how IDO suppresses the antiglioma immune response and/or response to glioma immunotherapy in young and older adult subjects (Fig. 4). Greater than 90% of human patient-resected GBM expresses IDO [64] and intratumoral IDO expression levels are directly increased by tumor-infiltrating T cells [64, 110] but unchanged by the effects of aging [31]. Glioma cell IDO increases intratumoral immunosuppressive regulatory T cells (Tregs; $CD4^+CD25^+FoxP3^+$) [86] and decreases overall survival regardless of treatment with a pharmacologic IDO enzyme inhibitor [87] or IDO metabolism by glioma cells [36, 123]. Advanced age is associated with increased nonmetabolic immunosuppressive IDO activity in the brain [31, 87, 94], which inhibits T-cell effector functions and decreases therapeutic efficacy against glioma. Senescent cells increase in the brain during advanced age and express the senescence-associated secretory phenotype (SASP), which in turn increases inflammation-inducible immunosuppressive IDO levels due to IDO promoter sensitivity to proinflammatory stimuli [119].

Concluding remarks

We have reviewed the role, expression pattern, and dynamic functionality of IDO in subjects with brain cancer many times (Fig. 5) [18, 105, 118, 119, 126]. Each iteration of new

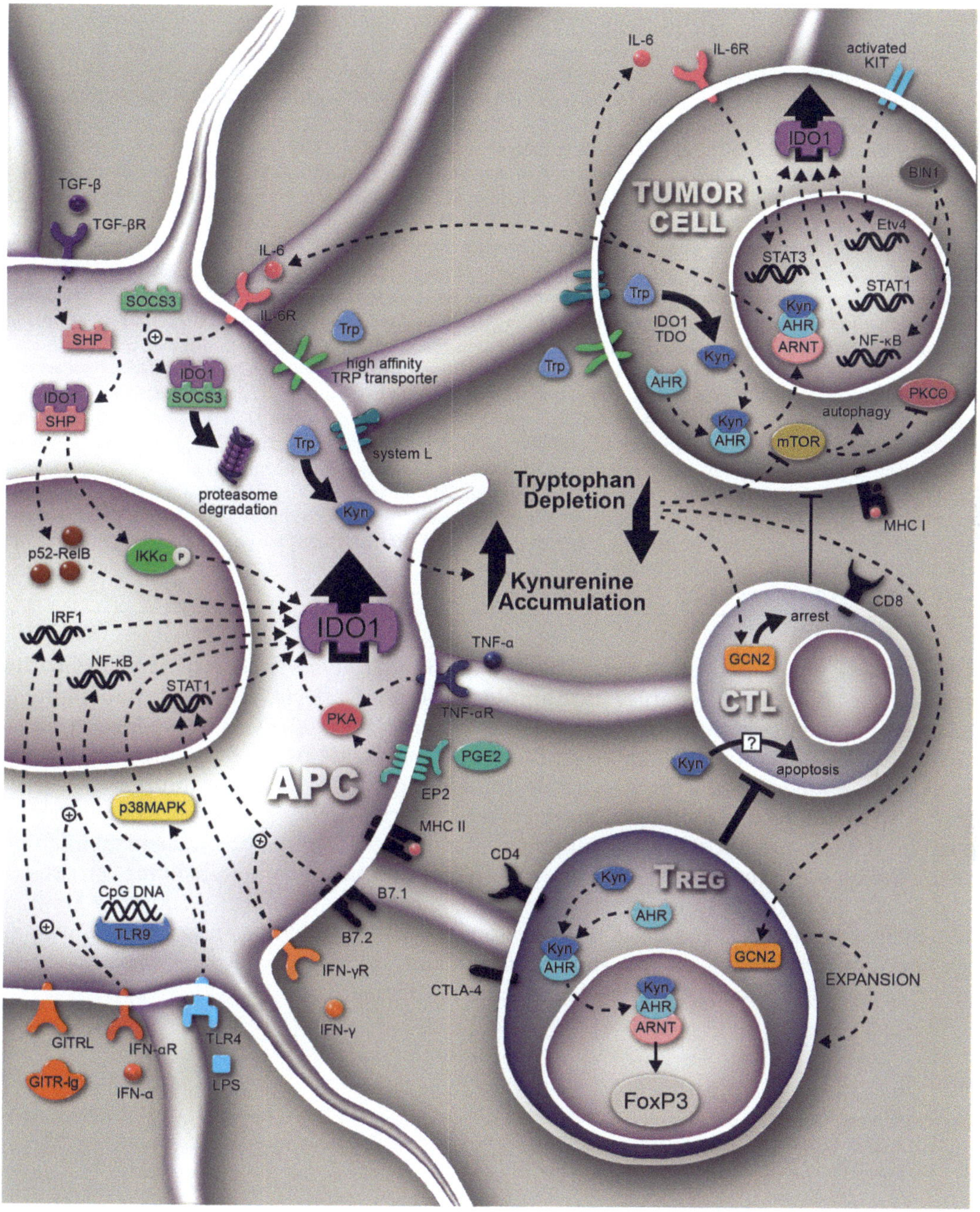

FIG. 5 (See the legend on next page)

FIG. 5, Signaling pathways associated with IDO and cancer [119]. The high expression of active IDO1 leads to a commensurately high rate of tryptophan conversion and depletion. This induces cell cycle arrest and/or anergy in the effector cytotoxic lymphocyte (CTL) compartment via the eIF2a kinase-dependent GCN2 pathway. Simultaneously, this mechanism also contributes to the activation/maturation of Treg in association with CTLA4-mediated CD80/CD86 co-inhibition. Kynurenine (Kyn) directly induces the apoptosis of CTL by an uncharacterized mechanism, while interacting with the aryl hydrocarbon receptor (AhR) in naïve $CD4^+$ T cells, resulting in the induction of $FoxP3^+$ iTreg. AhR interacts with the aryl hydrocarbon receptor nuclear translocator (ARNT) to mediate the specific transcriptional programming. Coincidently, IDO1 nonenzymatically enforces immunosuppression through two intrinsic immunoreceptor tyrosine-based inhibitory motifs (ITIM) in antigen-presenting cells (APC). TGF-β signaling results in the phosphorylation of the IDO1 ITIM, triggering noncanonical NF-κB activation and phosphorylation of IKKα, followed by nuclear translocation of the NF-κB subunits, p52 and RelB and autocrine reinforcement of IDO1 and TGF-β expression. The high-affinity Trp transporter is expressed by APC and tumor cells, with the majority of agonists leading to IDO1 activity demonstrated in APC. Similarly, the oncogene, c-KIT, and tumor-suppressor gene, Bin1, as well as the IL-6/AhR/STAT-3 signaling loop, have also been shown to impact the regulation of IDO1 in tumor cells [37, 39, 47, 50, 51, 62, 84, 124, 125]. Notably, in the presence of IDO1/IDO2/TDO, Kyn accumulation simultaneously contributes to Treg activation, promoting the disabling and/or apoptosis of CTL, thereby supporting tumor outgrowth by virtue of an unproductive antitumor response. *TGF-βR*, transforming growth factor-beta receptor; *IL-6R*, interleukin-6 receptor; *IRF1*, interferon response factor 1; *GITRL*, glucocorticoid-induced TNFR-related protein ligand; *IFN-aR*, interferon alpha receptor; *TLR4*, toll-like receptor 4; *LPS*, lipopolysaccharide; *MHC I/II*, major histocompatibility complex I/II; *EP2*, prostaglandin receptor E2; *PGE2*, prostaglandin E2; *TNF-αR*, tumor necrosis factor alpha; *SHP*, SH2 domain containing protein tyrosine phosphatase; *SOCS3*, suppressor of cytokine signaling 3; *mTOR*, mammalian target of rapamycin; *Etv4*, ETS translocation variant 4; *STAT*, signal transducer and activator of transcription; *PKCθ*, protein kinase C theta; *Bin1*, Myc box-dependent-interacting protein 1; *GCN2*, general control nonderepressible 2; *PKA*, protein kinase A.

understanding and perspective has provided additional layers of complexity that further questions the ability to fully target its immunosuppressive effects. The most challenging aspect of explaining its immunoregulatory effects is its multimodal levels of expression and activity including intratumoral, outside of the tumor but inside of the brain during advanced age, as well as outside of the brain in the draining lymph nodes during immunological priming. Additionally, its metabolic and nonmetabolic effects convey a pleiotropic functionality. To inhibit the full aspects of IDO in subjects with GBM, a drug that has wide bioavailability, crosses the blood-brain barrier, accumulates in the normal brain parenchyma, crosses the blood-tumor barrier, accumulates intratumorally, and one that neutralizes both IDO enzyme and nonenzyme effects is required. Our group is worked on such an endeavor and we have high hopes of providing a second-generation IDO inhibitory technology for improving immunotherapeutic effectiveness in GBM patients in the near future.

Funding

This work was supported by NIH Grants R01 NS097851 (D.A.W.), P50 CA221747 Project 2 (D.A.W.), K02 AG068617 (D.A.W.), T32 CA070085 (E.L.), BrainUp Grant 2136 (D.A.W.), the Gail Boyter Magness (GBM) Foundation (D.A.W.), and the Grace Giving Foundation (D.A.W.).

References

[1] Fontenot JD, Gavin MA, Rudensky AY. Foxp3 programs the development and function of CD4+CD25+ regulatory T cells. Nat Immunol 2003;4:330–6.

[2] Annunziato F, Cosmi L, Liotta F, Lazzeri E, Manetti R, Vanini V, Romagnani P, Maggi E, Romagnani S. Phenotype, localization, and mechanism of suppression of CD4(+)CD25(+) human thymocytes. J Exp Med 2002;196:379–87.

[3] Bluestone JA, Abbas AK. Natural versus adaptive regulatory T cells. Nat Rev Immunol 2003;3:253–7.

[4] Curiel TJ, Coukos G, Zou L, Alvarez X, Cheng P, Mottram P, Evdemon-Hogan M, Conejo-Garcia JR, Zhang L, Burow M, Zhu Y, Wei S, Kryczek I, Daniel B, Gordon A, Myers L, Lackner A, Disis ML, Knutson KL, Chen L, Zou W. Specific recruitment of regulatory T cells in ovarian carcinoma fosters immune privilege and predicts reduced survival. Nat Med 2004;10:942–9.

[5] El Andaloussi A, Lesniak MS. An increase in CD4+CD25+FOXP3+ regulatory T cells in tumor-infiltrating lymphocytes of human glioblastoma multiforme. Neuro-Oncology 2006;8:234–43.

[6] Konishi J, Yamazaki K, Azuma M, Kinoshita I, Dosaka-Akita H, Nishimura M. B7-H1 expression on non-small cell lung cancer cells and its relationship with tumor-infiltrating lymphocytes and their PD-1 expression. Clin Cancer Res 2004;10:5094–100.

[7] Dong H, Strome SE, Salomao DR, Tamura H, Hirano F, Flies DB, Roche PC, Lu J, Zhu G, Tamada K, Lennon VA, Celis E, Chen L. Tumor-associated B7-H1 promotes T-cell apoptosis: a potential mechanism of immune evasion. Nat Med 2002;8:793–800.

[8] Bloch O, Crane CA, Kaur R, Safaee M, Rutkowski MJ, Parsa AT. Gliomas promote immunosuppression through induction of B7-H1 expression in tumor-associated macrophages. Clin Cancer Res 2013;19:3165–75.

[9] Topalian SL, Hodi FS, Brahmer JR, Gettinger SN, Smith DC, McDermott DF, Powderly JD, Carvajal RD, Sosman JA, Atkins MB, Leming PD, Spigel DR, Antonia SJ, Horn L, Drake CG, Pardoll DM, Chen L, Sharfman WH, Anders RA, Taube JM, McMiller TL, Xu H, Korman AJ, Jure-Kunkel M, Agrawal S, McDonald D, Kollia GD, Gupta A, Wigginton JM, Sznol M. Safety, activity, and immune correlates of anti-PD-1 antibody in cancer. N Engl J Med 2012;366:2443–54.

[10] Brahmer JR, Tykodi SS, Chow LQ, Hwu WJ, Topalian SL, Hwu P, Drake CG, Camacho LH, Kauh J, Odunsi K, Pitot HC, Hamid O, Bhatia S, Martins R, Eaton K, Chen S, Salay TM, Alaparthy S, Grosso JF, Korman AJ, Parker SM, Agrawal S, Goldberg SM, Pardoll DM, Gupta A, Wigginton JM. Safety and activity of anti-PD-L1 antibody in patients with advanced cancer. N Engl J Med 2012;366:2455–65.

[11] Hodi FS, O'Day SJ, McDermott DF, Weber RW, Sosman JA, Haanen JB, Gonzalez R, Robert C, Schadendorf D, Hassel JC, Akerley W, van den Eertwegh AJ, Lutzky J, Lorigan P, Vaubel JM, Linette GP, Hogg D, Ottensmeier CH, Lebbe C, Peschel C, Quirt I, Clark JI, Wolchok JD, Weber JS, Tian J, Yellin MJ, Nichol GM, Hoos A, Urba WJ. Improved survival with ipilimumab in patients with metastatic melanoma. N Engl J Med 2010;363:711–23.

[12] Wolchok JD, Kluger H, Callahan MK, Postow MA, Rizvi NA, Lesokhin AM, Segal NH, Ariyan CE, Gordon R-A, Reed K, Burke MM, Caldwell A, Kronenberg SA, Agunwamba BU, Zhang X, Lowy I, Inzunza HD, Feely W, Horak CE, Hong Q, Korman AJ, Wigginton JM, Gupta A, Sznol M. Nivolumab plus ipilimumab in advanced melanoma. N Engl J Med 2013;369:122–33.

[13] Larkin J, Chiarion-Sileni V, Gonzalez R, Grob JJ, Cowey CL, Lao CD, Schadendorf D, Dummer R, Smylie M, Rutkowski P, Ferrucci PF, Hill A, Wagstaff J, Carlino MS, Haanen JB, Maio M, Marquez-Rodas I, McArthur GA, Ascierto PA, Long GV, Callahan MK, Postow MA, Grossmann K, Sznol M, Dreno B, Bastholt L, Yang A, Rollin LM, Horak C, Hodi FS, Wolchok JD. Combined nivolumab and ipilimumab or monotherapy in untreated melanoma. N Engl J Med 2015;373:23–34.

[14] Duraiswamy J, Kaluza KM, Freeman GJ, Coukos G. Dual blockade of PD-1 and CTLA-4 combined with tumor vaccine effectively restores T-cell rejection function in tumors. Cancer Res 2013;73:3591–603.

[15] Wainwright DA, Chang AL, Dey M, Balyasnikova IV, Kim C, Tobias AL, Cheng Y, Kim J, Zhang L, Qiao J, Han Y, Lesniak MS. Durable therapeutic efficacy utilizing combinatorial blockade against IDO, CTLA-4 and PD-L1 in mice with brain tumors. Clin Cancer Res 2014;20(20):5290–301.

[16] Weber J. Immune checkpoint proteins: a new therapeutic paradigm for cancer—preclinical background: CTLA-4 and PD-1 blockade. Semin Oncol 2010;37:430–9.

[17] Badawy AA-B, Guillemin G. The plasma [kynurenine]/[tryptophan] ratio and indoleamine 2,3-dioxygenase: time for appraisal. Int J Tryptop Res 2019;12, 1178646919868978.

[18] Zhai L, Ladomersky E, Lenzen A, Nguyen B, Patel R, Lauing KL, Wu M, Wainwright DA. IDO1 in cancer: a Gemini of immune checkpoints. Cell Mol Immunol 2018;15:447–57.

[19] Grohmann U, Fallarino F, Puccetti P. Tolerance, DCs and tryptophan: much ado about IDO. Trends Immunol 2003;24:242–8.

[20] Batabyal D, Yeh SR. Human tryptophan dioxygenase: a comparison to indoleamine 2,3-dioxygenase. J Am Chem Soc 2007;129:15690–701.
[21] Basran J, Rafice SA, Chauhan N, Efimov I, Cheesman MR, Ghamsari L, Raven EL. A kinetic, spectroscopic, and redox study of human tryptophan 2,3-dioxygenase. Biochemistry 2008;47:4752–60.
[22] Pantouris G, Serys M, Yuasa HJ, Ball HJ, Mowat CG. Human indoleamine 2,3-dioxygenase-2 has substrate specificity and inhibition characteristics distinct from those of indoleamine 2,3-dioxygenase-1. Amino Acids 2014;46:2155–63.
[23] Taylor MW, Feng GS. Relationship between interferon-gamma, indoleamine 2,3-dioxygenase, and tryptophan catabolism. FASEB J 1991;5:2516–22.
[24] Mellor AL, Baban B, Chandler PR, Manlapat A, Kahler DJ, Munn DH. Cutting edge: CpG oligonucleotides induce splenic CD19+ dendritic cells to acquire potent indoleamine 2,3-dioxygenase-dependent T cell regulatory functions via IFN type 1 signaling. J Immunol 2005;175:5601–5.
[25] Fallarino F, Volpi C, Zelante T, Vacca C, Calvitti M, Fioretti MC, Puccetti P, Romani L, Grohmann U. IDO mediates TLR9-driven protection from experimental autoimmune diabetes. J Immunol 2009;183:6303–12.
[26] Fujigaki H, Saito K, Fujigaki S, Takemura M, Sudo K, Ishiguro H, Seishima M. The signal transducer and activator of transcription 1α and interferon regulatory factor 1 are not essential for the induction of indoleamine 2,3-dioxygenase by lipopolysaccharide: involvement of p38 mitogen-activated protein kinase and nuclear factor-κB pathways, and synergistic effect of several proinflammatory cytokines. J Biochem 2006;139:655–62.
[27] Balachandran VP, Cavnar MJ, Zeng S, Bamboat ZM, Ocuin LM, Obaid H, Sorenson EC, Popow R, Ariyan C, Rossi F, Besmer P, Guo T, Antonescu CR, Taguchi T, Yuan J, Wolchok JD, Allison JP, DeMatteo RP. Imatinib potentiates antitumor T cell responses in gastrointestinal stromal tumor through the inhibition of Ido. Nat Med 2011;17:1094–100.
[28] Grohmann U, Orabona C, Fallarino F, Vacca C, Calcinaro F, Falorni A, Candeloro P, Belladonna ML, Bianchi R, Fioretti MC, Puccetti P. CTLA-4-Ig regulates tryptophan catabolism in vivo. Nat Immunol 2002;3:1097–101.
[29] Holtzhausen A, Zhao F, Evans KS, Tsutsui M, Orabona C, Tyler DS, Hanks BA. Melanoma-derived Wnt5a promotes local dendritic-cell expression of IDO and immunotolerance: opportunities for pharmacologic enhancement of immunotherapy. Cancer Immunol Res 2015;3:1082–95.
[30] Theate I, van Baren N, Pilotte L, Moulin P, Larrieu P, Renauld JC, Herve C, Gutierrez-Roelens I, Marbaix E, Sempoux C, Van den Eynde BJ. Extensive profiling of the expression of the indoleamine 2,3-dioxygenase 1 protein in normal and tumoral human tissues. Cancer Immunol Res 2015;3:161–72.
[31] Ladomersky E, Scholtens DM, Kocherginsky M, Hibler EA, Bartom ET, Otto-Meyer S, Zhai L, Lauing KL, Choi J, Sosman JA, Wu JD, Zhang B, Lukas RV, Wainwright DA. The coincidence between increasing age, immunosuppression, and the incidence of patients with glioblastoma. Front Pharmacol 2019;10:200.
[32] Litzenburger UM, Opitz CA, Sahm F, Rauschenbach KJ, Trump S, Winter M, Ott M, Ochs K, Lutz C, Liu X, Anastasov N, Lehmann I, Höfer T, von Deimling A, Wick W, Platten M. Constitutive IDO expression in human cancer is sustained by an autocrine signaling loop involving IL-6, STAT3 and the AHR. Oncotarget 2014.
[33] Li Q, Harden JL, Anderson CD, Egilmez NK. Tolerogenic phenotype of IFN-gamma-induced IDO+ dendritic cells is maintained via an autocrine IDO-kynurenine/AhR-IDO loop. J Immunol 2016;197:962–70.
[34] Hennequart M, Pilotte L, Cane S, Hoffmann D, Stroobant V, Plaen E, Van den Eynde BJ. Constitutive IDO1 expression in human tumors is driven by cyclooxygenase-2 and mediates intrinsic immune resistance. Cancer Immunol Res 2017;5:695–709.
[35] Miyazaki T, Moritake K, Yamada K, Hara N, Osago H, Shibata T, Akiyama Y, Tsuchiya M. Indoleamine 2,3-dioxygenase as a new target for malignant glioma therapy. Laboratory investigation. J Neurosurg 2009;111:230–7.
[36] Zhai L, Ladomersky E, Dostal CR, Lauing KL, Swoap K, Billingham LK, Gritsina G, Wu M, McCusker RH, Binder DC, Wainwright DA. Non-tumor cell IDO1 predominantly contributes to enzyme activity and response to CTLA-4/PD-L1 inhibition in mouse glioblastoma. Brain Behav Immun 2017;62:24–9.
[37] Munn DH, Sharma MD, Baban B, Harding HP, Zhang Y, Ron D, Mellor AL. GCN2 kinase in T cells mediates proliferative arrest and anergy induction in response to indoleamine 2,3-dioxygenase. Immunity 2005;22:633–42.
[38] Metz R, Rust S, Duhadaway JB, Mautino MR, Munn DH, Vahanian NN, Link CJ, Prendergast GC. IDO inhibits a tryptophan sufficiency signal that stimulates mTOR: a novel IDO effector pathway targeted by D-1-methyl-tryptophan. Onco Targets Ther 2012;1:1460–8.
[39] Mezrich JD, Fechner JH, Zhang X, Johnson BP, Burlingham WJ, Bradfield CA. An interaction between kynurenine and the aryl hydrocarbon receptor can generate regulatory T cells. J Immunol 2010;185:3190–8.

[40] Stockinger B, Di Meglio P, Gialitakis M, Duarte JH. The aryl hydrocarbon receptor: multitasking in the immune system. Annu Rev Immunol 2014;32:403–32.
[41] Quintana FJ, Basso AS, Iglesias AH, Korn T, Farez MF, Bettelli E, Caccamo M, Oukka M, Weiner HL. Control of Treg and TH17 cell differentiation by the aryl hydrocarbon receptor. Nature 2008;453:65–71.
[42] Marshall NB, Kerkvliet NI. Dioxin and immune regulation: emerging role of aryl hydrocarbon receptor in the generation of regulatory T cells. Ann N Y Acad Sci 2010;1183:25–37.
[43] DiNatale BC, Murray IA, Schroeder JC, Flaveny CA, Lahoti TS, Laurenzana EM, Omiecinski CJ, Perdew GH. Kynurenic acid is a potent endogenous aryl hydrocarbon receptor ligand that synergistically induces interleukin-6 in the presence of inflammatory Signaling. Toxicol Sci 2010;115:89–97.
[44] Novikov O, Wang Z, Stanford EA, Parks AJ, Ramirez-Cardenas A, Landesman E, Laklouk I, Sarita-Reyes C, Gusenleitner D, Li A, Monti S, Manteiga S, Lee K, Sherr DH. An aryl hydrocarbon receptor-mediated amplification loop that enforces cell migration in ER-/PR-/Her2- human breast cancer cells. Mol Pharmacol 2016;90:674–88.
[45] Lowe MM, Mold JE, Kanwar B, Huang Y, Louie A, Pollastri MP, Wang C, Patel G, Franks DG, Schlezinger J, Sherr DH, Silverstone AE, Hahn ME, McCune JM. Identification of cinnabarinic acid as a novel endogenous aryl hydrocarbon receptor ligand that drives IL-22 production. PLoS ONE 2014;9, e87877.
[46] Terness P, Bauer TM, Röse L, Dufter C, Watzlik A, Simon H, Opelz G. Inhibition of allogeneic T cell proliferation by indoleamine 2,3-dioxygenase–expressing dendritic cells: mediation of suppression by tryptophan metabolites. J Exp Med 2002;196:447–57.
[47] Fallarino F, Grohmann U, Vacca C, Bianchi R, Orabona C, Spreca A, Fioretti MC, Puccetti P. T cell apoptosis by tryptophan catabolism. Cell Death Differ 2002;9:1069–77.
[48] Frumento G, Rotondo R, Tonetti M, Damonte G, Benatti U, Ferrara GB. Tryptophan-derived catabolites are responsible for inhibition of T and natural killer cell proliferation induced by indoleamine 2,3-dioxygenase. J Exp Med 2002;196:459–68.
[49] Segawa H, Fukasawa Y, Miyamoto K-i, Takeda E, Endou H, Kanai Y. Identification and functional characterization of a Na+-independent neutral amino acid transporter with broad substrate selectivity. J Biol Chem 1999;274:19745–51.
[50] Seymour RL, Ganapathy V, Mellor AL, Munn DH. A high-affinity, tryptophan-selective amino acid transport system in human macrophages. J Leukoc Biol 2006;80:1320–7.
[51] Silk JD, Lakhal S, Laynes R, Vallius L, Karydis I, Marcea C, Boyd CAR, Cerundolo V. IDO induces expression of a novel tryptophan transporter in mouse and human tumor cells. J Immunol 2011;187:1617–25.
[52] Schrocksnadel K, Wirleitner B, Winkler C, Fuchs D. Monitoring tryptophan metabolism in chronic immune activation. Clin Chim Acta 2006;364:82–90.
[53] Huang A, Fuchs D, Widner B, Glover C, Henderson DC, Allen-Mersh TG. Serum tryptophan decrease correlates with immune activation and impaired quality of life in colorectal cancer. Br J Cancer 2002;86:1691–6.
[54] Li M, Bolduc AR, Hoda MN, Gamble DN, Dolisca SB, Bolduc AK, Hoang K, Ashley C, McCall D, Rojiani AM, Maria BL, Rixe O, MacDonald TJ, Heeger PS, Mellor AL, Munn DH, Johnson TS. The indoleamine 2,3-dioxygenase pathway controls complement-dependent enhancement of chemo-radiation therapy against murine glioblastoma. J Immunother Cancer 2014;2:21.
[55] Zhai L, Dey M, Lauing KL, Gritsina G, Kaur R, Lukas RV, Nicholas MK, Rademaker AW, Dostal CR, McCusker RH, Raizer JJ, Parsa AT, Bloch O, Wainwright DA. The kynurenine to tryptophan ratio as a prognostic tool for glioblastoma patients enrolling in immunotherapy. J Clin Neurosci 2015;22:5.
[56] Adams S, Teo C, McDonald KL, Zinger A, Bustamante S, Lim CK, Sundaram G, Braidy N, Brew BJ, Guillemin GJ. Involvement of the kynurenine pathway in human glioma pathophysiology. PLoS ONE 2014;9, e112945.
[57] Takenaka MC, Gabriely G, Rothhammer V, Mascanfroni ID, Wheeler MA, Chao CC, Gutierrez-Vazquez C, Kenison J, Tjon EC, Barroso A, Vandeventer T, de Lima KA, Rothweiler S, Mayo L, Ghannam S, Zandee S, Healy L, Sherr D, Farez MF, Prat A, Antel J, Reardon DA, Zhang H, Robson SC, Getz G, Weiner HL, Quintana FJ. Control of tumor-associated macrophages and T cells in glioblastoma via AHR and CD39. Nat Neurosci 2019;22:729–40.
[58] Nguyen NT, Kimura A, Nakahama T, Chinen I, Masuda K, Nohara K, Fujii-Kuriyama Y, Kishimoto T. Aryl hydrocarbon receptor negatively regulates dendritic cell immunogenicity via a kynurenine-dependent mechanism. Proc Natl Acad Sci 2010;107:19961–6.
[59] Wainwright DA, Sengupta S, Han Y, Lesniak MS. Thymus-derived rather than tumor-induced regulatory T cells predominate in brain tumors. Neuro-Oncology 2011;13:1308–23.

[60] Malchow S, Leventhal DS, Nishi S, Fischer BI, Shen L, Paner GP, Amit AS, Kang C, Geddes JE, Allison JP, Socci ND, Savage PA. Aire-dependent thymic development of tumor-associated regulatory T cells. Science 2013;339:1219–24.

[61] Ye J, Qiu J, Bostick JW, Ueda A, Schjerven H, Li S, Jobin C, Chen ZE, Zhou L. The aryl hydrocarbon receptor preferentially marks and promotes gut regulatory T cells. Cell Rep 2017;21:2277–90.

[62] Sharma MD, Baban B, Chandler P, Hou D-Y, Singh N, Yagita H, Azuma M, Blazar BR, Mellor AL, Munn DH. Plasmacytoid dendritic cells from mouse tumor-draining lymph nodes directly activate mature Tregs via indoleamine 2,3-dioxygenase. J Clin Invest 2007;117:2570–82.

[63] Sharma MD, Shinde R, McGaha TL, Huang L, Holmgaard RB, Wolchok JD, Mautino MR, Celis E, Sharpe AH, Francisco LM, Powell JD, Yagita H, Mellor AL, Blazar BR, Munn DH. The PTEN pathway in Tregs is a critical driver of the suppressive tumor microenvironment. Sci Adv 2015;1, e1500845.

[64] Zhai L, Ladomersky E, Lauing KL, Wu M, Genet M, Gritsina G, Gyorffy B, Brastianos PK, Binder DC, Sosman JA, Giles FJ, James CD, Horbinski C, Stupp R, Wainwright DA. Infiltrating T cells increase IDO1 expression in glioblastoma and contribute to decreased patient survival. Clin Cancer Res 2017;23:6650–60.

[65] Uyttenhove C, Pilotte L, Theate I, Stroobant V, Colau D, Parmentier N, Boon T, Van den Eynde BJ. Evidence for a tumoral immune resistance mechanism based on tryptophan degradation by indoleamine 2,3-dioxygenase. Nat Med 2003;9:1269–74.

[66] Riesenberg R, Weiler C, Spring O, Eder M, Buchner A, Popp T, Castro M, Kammerer R, Takikawa O, Hatz RA, Stief CG, Hofstetter A, Zimmermann W. Expression of indoleamine 2,3-dioxygenase in tumor endothelial cells correlates with long-term survival of patients with renal cell carcinoma. Clin Cancer Res 2007;13:6993–7002.

[67] Ishio T, Goto S, Tahara K, Tone S, Kawano K, Kitano S. Immunoactivative role of indoleamine 2,3-dioxygenase in human hepatocellular carcinoma. J Gastroenterol Hepatol 2004;19:319–26.

[68] Pan K, Wang H, Chen MS, Zhang HK, Weng DS, Zhou J, Huang W, Li JJ, Song HF, Xia JC. Expression and prognosis role of indoleamine 2,3-dioxygenase in hepatocellular carcinoma. J Cancer Res Clin Oncol 2008;134:1247–53.

[69] Piras F, Colombari R, Minerba L, Murtas D, Floris C, Maxia C, Corbu A, Perra MT, Sirigu P. The predictive value of CD8, CD4, CD68, and human leukocyte antigen-D-related cells in the prognosis of cutaneous malignant melanoma with vertical growth phase. Cancer 2005;104:1246–54.

[70] Tang D, Yue L, Yao R, Zhou L, Yang Y, Lu L, Gao W. P53 prevent tumor invasion and metastasis by down-regulating IDO in lung cancer. Oncotarget 2017;8:54548–57.

[71] Levina V, Su Y, Gorelik E. Immunological and nonimmunological effects of indoleamine 2,3-dioxygenase on breast tumor growth and spontaneous metastasis formation. Clin Dev Immunol 2012;2012:173029.

[72] Smith C, Chang MY, Parker KH, Beury DW, DuHadaway JB, Flick HE, Boulden J, Sutanto-Ward E, Soler AP, Laury-Kleintop LD, Mandik-Nayak L, Metz R, Ostrand-Rosenberg S, Prendergast GC, Muller AJ. IDO is a nodal pathogenic driver of lung cancer and metastasis development. Cancer Discov 2012;2:722–35.

[73] Brandacher G, Perathoner A, Ladurner R, Schneeberger S, Obrist P, Winkler C, Werner ER, Werner-Felmayer G, Weiss HG, Gobel G, Margreiter R, Konigsrainer A, Fuchs D, Amberger A. Prognostic value of indoleamine 2,3-dioxygenase expression in colorectal cancer: effect on tumor-infiltrating T cells. Clin Cancer Res 2006;12:1144–51.

[74] Ino K, Yamamoto E, Shibata K, Kajiyama H, Yoshida N, Terauchi M, Nawa A, Nagasaka T, Takikawa O, Kikkawa F. Inverse correlation between tumoral indoleamine 2,3-dioxygenase expression and tumor-infiltrating lymphocytes in endometrial cancer: its association with disease progression and survival. Clin Cancer Res 2008;14:2310–7.

[75] Maleki Vareki S, Rytelewski M, Figueredo R, Chen D, Ferguson PJ, Vincent M, Min W, Zheng X, Koropatnick J. Indoleamine 2,3-dioxygenase mediates immune-independent human tumor cell resistance to olaparib, gamma radiation, and cisplatin. Oncotarget 2014;5:2778–91.

[76] Tummala KS, Gomes AL, Yilmaz M, Grana O, Bakiri L, Ruppen I, Ximenez-Embun P, Sheshappanavar V, Rodriguez-Justo M, Pisano DG, Wagner EF, Djouder N. Inhibition of de novo NAD(+) synthesis by oncogenic URI causes liver tumorigenesis through DNA damage. Cancer Cell 2014;26:826–39.

[77] Wang Y, Liu H, McKenzie G, Witting PK, Stasch JP, Hahn M, Changsirivathanathamrong D, Wu BJ, Ball HJ, Thomas SR, Kapoor V, Celermajer DS, Mellor AL, Keaney Jr JF, Hunt NH, Stocker R. Kynurenine is an endothelium-derived relaxing factor produced during inflammation. Nat Med 2010;16:279–85.

[78] Sakakibara K, Feng GG, Li J, Akahori T, Yasuda Y, Nakamura E, Hatakeyama N, Fujiwara Y, Kinoshita H. Kynurenine causes vasodilation and hypotension induced by activation of KCNQ-encoded voltage-dependent K(+) channels. J Pharmacol Sci 2015;129:31–7.
[79] Fazio F, Carrizzo A, Lionetto L, Damato A, Capocci L, Ambrosio M, Battaglia G, Bruno V, Madonna M, Simmaco M, Nicoletti F, Vecchione C. Vasorelaxing action of the kynurenine metabolite, xanthurenic acid: the missing Link in endotoxin-induced hypotension? Front Pharmacol 2017;8:214.
[80] Stanley CP, Maghzal GJ, Ayer A, Talib J, Giltrap AM, Shengule S, Wolhuter K, Wang Y, Chadha P, Suarna C, Prysyazhna O, Scotcher J, Dunn LL, Prado FM, Nguyen N, Odiba JO, Baell JB, Stasch JP, Yamamoto Y, Di Mascio P, Eaton P, Payne RJ, Stocker R. Singlet molecular oxygen regulates vascular tone and blood pressure in inflammation. Nature 2019;566:548–52.
[81] Changsirivathanathamrong D, Wang Y, Rajbhandari D, Maghzal GJ, Mak WM, Woolfe C, Duflou J, Gebski V, dos Remedios CG, Celermajer DS, Stocker R. Tryptophan metabolism to kynurenine is a potential novel contributor to hypotension in human sepsis. Crit Care Med 2011;39:2678–83.
[82] Mondal A, Smith C, DuHadaway JB, Sutanto-Ward E, Prendergast GC, Bravo-Nuevo A, Muller AJ. IDO1 is an integral mediator of inflammatory neovascularization. EBioMedicine 2016;14:74–82.
[83] Prendergast GC, Mondal A, Dey S, Laury-Kleintop LD, Muller AJ. Inflammatory reprogramming with IDO1 inhibitors: turning immunologically unresponsive 'Cold' tumors 'Hot'. Trends cancer 2018;4:38–58.
[84] Pallotta MT, Orabona C, Volpi C, Vacca C, Belladonna ML, Bianchi R, Servillo G, Brunacci C, Calvitti M, Bicciato S, Mazza EMC, Boon L, Grassi F, Fioretti MC, Fallarino F, Puccetti P, Grohmann U. Indoleamine 2,3-dioxygenase is a signaling protein in long-term tolerance by dendritic cells. Nat Immunol 2011;12:870–8.
[85] Liu X, Shin N, Koblish HK, Yang G, Wang Q, Wang K, Leffet L, Hansbury MJ, Thomas B, Rupar M, Waeltz P, Bowman KJ, Polam P, Sparks RB, Yue EW, Li Y, Wynn R, Fridman JS, Burn TC, Combs AP, Newton RC, Scherle PA. Selective inhibition of IDO1 effectively regulates mediators of antitumor immunity. Blood 2010;115:3520–30.
[86] Wainwright DA, Balyasnikova IV, Chang AL, Ahmed AU, Moon KS, Auffinger B, Tobias AL, Han Y, Lesniak MS. IDO expression in brain tumors increases the recruitment of regulatory T cells and negatively impacts survival. Clin Cancer Res 2012;18:6110–21.
[87] Ladomersky E, Zhai L, Lenzen A, Lauing KL, Qian J, Scholtens DM, Gritsina G, Sun X, Liu Y, Yu F, Gong W, Liu Y, Jiang B, Tang T, Patel R, Platanias LC, James CD, Stupp R, Lukas RV, Binder DC, Wainwright DA. IDO1 inhibition synergizes with radiation and PD-1 blockade to durably increase survival against advanced glioblastoma. Clin Cancer Res 2018;24:2559–73.
[88] Frick B, Schroecksnadel K, Neurauter G, Leblhuber F, Fuchs D. Increasing production of homocysteine and neopterin and degradation of tryptophan with older age. Clin Biochem 2004;37:684–7.
[89] Pertovaara M, Raitala A, Lehtimaki T, Karhunen PJ, Oja SS, Jylha M, Hervonen A, Hurme M. Indoleamine 2,3-dioxygenase activity in nonagenarians is markedly increased and predicts mortality. Mech Ageing Dev 2006;127:497–9.
[90] Marcos-Perez D, Sanchez-Flores M, Maseda A, Lorenzo-Lopez L, Millan-Calenti JC, Strasser B, Gostner JM, Fuchs D, Pasaro E, Valdiglesias V, Laffon B. Frailty status in older adults is related to alterations in indoleamine 2,3-dioxygenase 1 and guanosine triphosphate cyclohydrolase I enzymatic pathways. J Am Med Dir Assoc 2017;18:1049–57.
[91] Sutphin GL, Backer G, Sheehan S, Bean S, Corban C, Liu T, Peters MJ, van Meurs JBJ, Murabito JM, Johnson AD, Korstanje R. Caenorhabditis elegans orthologs of human genes differentially expressed with age are enriched for determinants of longevity. Aging Cell 2017;16:672–82.
[92] Apalset EM, Gjesdal CG, Ueland PM, Midttun O, Ulvik A, Eide GE, Meyer K, Tell GS. Interferon (IFN)-gamma-mediated inflammation and the kynurenine pathway in relation to bone mineral density: the Hordaland Health Study. Clin Exp Immunol 2014;176:452–60.
[93] Reyes Ocampo J, Lugo Huitron R, Gonzalez-Esquivel D, Ugalde-Muniz P, Jimenez-Anguiano A, Pineda B, Pedraza-Chaverri J, Rios C, Perez de la Cruz V. Kynurenines with neuroactive and redox properties: relevance to aging and brain diseases. Oxidative Med Cell Longev 2014;2014:646909.
[94] Ladomersky E, Zhai L, Gritsina G, Genet M, Lauing KL, Wu M, James CD, Wainwright DA. Advanced age negatively impacts survival in an experimental brain tumor model. Neurosci Lett 2016;630:203–8.
[95] Refaey ME, McGee-Lawrence ME, Fulzele S, Kennedy EJ, Bollag WB, Elsalanty M, Zhong Q, Ding KH, Bendzunas NG, Shi XM, Xu J, Hill WD, Johnson MH, Hunter M, Pierce JL, Yu K, Hamrick MW, Isales CM. Kynurenine, a tryptophan metabolite that accumulates with age, induces bone loss. J Bone Miner Res 2017;32:2182–93.

[96] Minhas PS, Liu L, Moon PK, Joshi AU, Dove C, Mhatre S, Contrepois K, Wang Q, Lee BA, Coronado M, Bernstein D, Snyder MP, Migaud M, Majeti R, Mochly-Rosen D, Rabinowitz JD, Andreasson KI. Macrophage de novo NAD(+) synthesis specifies immune function in aging and inflammation. Nat Immunol 2019;20:50–63.

[97] Eckers A, Jakob S, Heiss C, Haarmann-Stemmann T, Goy C, Brinkmann V, Cortese-Krott MM, Sansone R, Esser C, Ale-Agha N, Altschmied J, Ventura N, Haendeler J. The aryl hydrocarbon receptor promotes aging phenotypes across species. Sci Rep 2016;6:19618.

[98] Ostrom QT, Gittleman H, Liao P, Vecchione-Koval T, Wolinsky Y, Kruchko C, Barnholtz-Sloan JS. CBTRUS statistical report: primary brain and other central nervous system tumors diagnosed in the United States in 2010-2014. Neuro-Oncology 2017;19:v1–v88.

[99] Mellor AL, Baban B, Chandler P, Marshall B, Jhaver K, Hansen A, Koni PA, Iwashima M, Munn DH. Cutting edge: induced indoleamine 2,3 dioxygenase expression in dendritic cell subsets suppresses T cell clonal expansion. J Immunol 2003;171:1652–5.

[100] Baban B, Chandler P, McCool D, Marshall B, Munn DH, Mellor AL. Indoleamine 2,3-dioxygenase expression is restricted to fetal trophoblast giant cells during murine gestation and is maternal genome specific. J Reprod Immunol 2004;61:67–77.

[101] Moreno ACR, Porchia B, Pagni RL, Souza PDC, Pegoraro R, Rodrigues KB, Barros TB, Aps L, de Araujo EF, Calich VLG, Ferreira LCS. The combined use of melatonin and an indoleamine 2,3-dioxygenase-1 inhibitor enhances vaccine-induced protective cellular immunity to HPV16-associated tumors. Front Immunol 2018;9:1914.

[102] Schafer CC, Wang Y, Hough KP, Sawant A, Grant SC, Thannickal VJ, Zmijewski J, Ponnazhagan S, Deshane JS. Indoleamine 2,3-dioxygenase regulates anti-tumor immunity in lung cancer by metabolic reprogramming of immune cells in the tumor microenvironment. Oncotarget 2016;7:75407–24.

[103] Takamatsu M, Hirata A, Ohtaki H, Hoshi M, Ando T, Ito H, Hatano Y, Tomita H, Kuno T, Saito K, Seishima M, Hara A. Inhibition of indoleamine 2,3-dioxygenase 1 expression alters immune response in colon tumor microenvironment in mice. Cancer Sci 2015;106:1008–15.

[104] Bishnupuri KS, Alvarado DM, Khouri AN, Shabsovich M, Chen B, Dieckgraefe BK, Ciorba MA. IDO1 and kynurenine pathway metabolites activate PI3K-Akt signaling in the neoplastic colon epithelium to promote cancer cell proliferation and inhibit apoptosis. Cancer Res 2019;79:1138–50.

[105] Zhai L, Bell A, Ladomersky E, Lauing KL, Bollu L, Sosman JA, Zhang B, Wu JD, Miller SD, Meeks JJ, Lukas RV, Wyatt E, Doglio L, Schiltz GE, McCusker RH, Wainwright DA. Immunosuppressive IDO in cancer: mechanisms of action, animal models, and targeting strategies. Front Immunol 2020;11:1185.

[106] Antonia SJ, Lopez-Martin JA, Bendell J, Ott PA, Taylor M, Eder JP, Jager D, Pietanza MC, Le DT, de Braud F, Morse MA, Ascierto PA, Horn L, Amin A, Pillai RN, Evans J, Chau I, Bono P, Atmaca A, Sharma P, Harbison CT, Lin CS, Christensen O, Calvo E. Nivolumab alone and nivolumab plus ipilimumab in recurrent small-cell lung cancer (CheckMate 032): a multicentre, open-label, phase 1/2 trial. Lancet Oncol 2016;17:883–95.

[107] Motzer RJ, Escudier B, McDermott DF, George S, Hammers HJ, Srinivas S, Tykodi SS, Sosman JA, Procopio G, Plimack ER, Castellano D, Choueiri TK, Gurney H, Donskov F, Bono P, Wagstaff J, Gauler TC, Ueda T, Tomita Y, Schutz FA, Kollmannsberger C, Larkin J, Ravaud A, Simon JS, Xu LA, Waxman IM, Sharma P, CheckMate I. Nivolumab versus everolimus in advanced renal-cell carcinoma. N Engl J Med 2015;373:1803–13.

[108] BMS. https://news.bms.com/news/details/2020/Bristol-Myers-Squibb-Announces-Update-on-Phase-3-CheckMate–548-Trial-Evaluating-Patients-with-Newly-Diagnosed-MGMT-Methylated-Glioblastoma-Multiforme/default.aspx.

[109] Reardon DA, Brandes AA, Omuro A, Mulholland P, Lim M, Wick A, Baehring J, Ahluwalia MS, Roth P, Bahr O, Phuphanich S, Sepulveda JM, De Souza P, Sahebjam S, Carleton M, Tatsuoka K, Taitt C, Zwirtes R, Sampson J, Weller M. Effect of nivolumab vs bevacizumab in patients with recurrent glioblastoma: the CheckMate 143 phase 3 randomized clinical trial. JAMA Oncol 2020;6(7):1003–10.

[110] O'Rourke DM, Nasrallah MP, Desai A, Melenhorst JJ, Mansfield K, Morrissette JJD, Martinez-Lage M, Brem S, Maloney E, Shen A, Isaacs R, Mohan S, Plesa G, Lacey SF, Navenot JM, Zheng Z, Levine BL, Okada H, June CH, Brogdon JL, Maus MV. A single dose of peripherally infused EGFRvIII-directed CAR T cells mediates antigen loss and induces adaptive resistance in patients with recurrent glioblastoma. Sci Transl Med 2017;9.

[111] Zhai L, Ladomersky E, Lauing KL, Wu M, Genet M, Gritsina G, Balaz G, Brastianos PK, Binder DC, Sosman JA, Giles FJ, James CD, Horbinski C, Stupp R, Wainwright DA. Infiltrating T cells increase IDO1 expression in glioblastoma and contribute to decreased patient survival. Clin Cancer Res 2017;23(21):6650–60.

[112] Zhai L, Ladomersky E, Lauing KL, Wu M, Scholtens DM, Savoor R, Zhang B, Wu JD, Horbinski C, Lukas RV, Binder DC, Wainwright DA. Commentary: preclinical efficacy of immune-checkpoint monotherapy does not

recapitulate corresponding biomarkers-based clinical predictions in glioblastoma by Garg et al. (2017). Oncoimmunology 2019;8, 1548242.

[113] Lohr J, Ratliff T, Huppertz A, Ge Y, Dictus C, Ahmadi R, Grau S, Hiraoka N, Eckstein V, Ecker RC, Korff T, von Deimling A, Unterberg A, Beckhove P, Herold-Mende C. Effector T-cell infiltration positively impacts survival of glioblastoma patients and is impaired by tumor-derived TGF-beta. Clin Cancer Res 2011;17:4296–308.

[114] El Andaloussi A, Han Y, Lesniak MS. Prolongation of survival following depletion of CD4+CD25+ regulatory T cells in mice with experimental brain tumors. J Neurosurg 2006;105:430–7.

[115] Mitchell DA, Cui X, Schmittling RJ, Sanchez-Perez L, Snyder DJ, Congdon KL, Archer GE, Desjardins A, Friedman AH, Friedman HS, Herndon 2nd JE, McLendon RE, Reardon DA, Vredenburgh JJ, Bigner DD, Sampson JH. Monoclonal antibody blockade of IL-2 receptor alpha during lymphopenia selectively depletes regulatory T cells in mice and humans. Blood 2011;118:3003–12.

[116] James CD. Stimulating anti-tumor immune response: the problem of regulatory T-cells. Neuro-Oncology 2011;13:1261.

[117] Wainwright DA, Nigam P, Thaci B, Dey M, Lesniak MS. Recent developments on immunotherapy for brain cancer. Expert Opin Emerg Drugs 2012;17(2):181–202.

[118] Wainwright DA, Dey M, Chang A, Lesniak MS. Targeting Tregs in malignant brain cancer: overcoming IDO. Front Immunol 2013;4:116.

[119] Zhai L, Spranger S, Binder DC, Gritsina G, Lauing KL, Giles FJ, Wainwright DA. Molecular pathways: targeting IDO1 and other tryptophan dioxygenases for cancer immunotherapy. Clin Cancer Res 2015;21:5427–33.

[120] Stupp R, Taillibert S, Kanner A, Read W, Steinberg DM, Lhermitte B, Toms S, Idbaih A, Ahluwalia MS, Fink K, Di Meco F, Lieberman F, Zhu JJ, Stragliotto G, Tran DD, Brem S, Hottinger AF, Kirson ED, Lavy-Shahaf G, Weinberg U, Kim CY, Paek SH, Nicholas G, Burna J, Hirte H, Weller M, Palti Y, Hegi ME, Ram Z. Effect of tumor-treating fields plus maintenance temozolomide vs maintenance temozolomide alone on survival in patients with glioblastoma: a randomized clinical trial. JAMA 2017;318:2306–16.

[121] Arvold ND, Reardon DA. Treatment options and outcomes for glioblastoma in the elderly patient. Clin Interv Aging 2014;9:357–67.

[122] Ladomersky E, Zhai L, Lauing KL, Bell A, Xu J, Kocherginsky M, Zhang B, Wu JD, Podojil JR, Platanias LC, Mochizuki AY, Prins RM, Kumthekar P, Raizer J, Dixit K, Lukas RV, Horbinski C, Wei M, Zhou C, Pawelec G, Campisi J, Grohmann U, Prendergast GC, Munn D, Wainwright DA. Advanced age increases immunosuppression in the brain and decreases immunotherapeutic efficacy in subjects with glioblastoma. Clin Cancer Res 2020;26(19):5232–45.

[123] Zhai L, Bell A, Qian J, Ladomersky E, Lauing KL, Wainwright DA. Tumor cell IDO1 increases intratumoral immunosuppressive Tregs independent of enzyme activity. J Immunol 2019;202:137.2.

[124] Muller AJ, DuHadaway JB, Donover PS, Sutanto-Ward E, Prendergast GC. Inhibition of indoleamine 2,3-dioxygenase, an immunoregulatory target of the cancer suppression gene Bin1, potentiates cancer chemotherapy. Nat Med 2005;11:312–9.

[125] Beischlag TV, Luis Morales J, Hollingshead BD, Perdew GH. The aryl hydrocarbon receptor complex and the control of gene expression. Crit Rev Eukaryot Gene Expr 2008;18:207–50.

[126] Zhai L, Lauing KL, Chang AL, Dey M, Qian J, Cheng Y, Lesniak MS, Wainwright DA. The role of IDO in brain tumor immunotherapy. J Neuro-Oncol 2015;123:395–403.

CHAPTER

9

Chemokine receptor antagonists as immunotherapy agents and adjuncts for glioblastoma

Adela Wu, Michael Jin, and Gordon Li

Department of Neurosurgery, Stanford University School of Medicine, Stanford, CA, United States

Abstract

Glioblastoma (GBM) is the most common primary central nervous system malignancy in adults and continues to carry a poor prognosis despite standard of care involving surgical resection and concomitant chemoradiotherapy. Recently, immunotherapy has become a rapidly evolving area of novel therapies for various cancers. For gliomas, antibodies including immune checkpoint inhibitors, such as anti-PD-1, and agents targeting tumor neovascularization were studied, but results from clinical trials show no survival benefit with immunotherapy in GBM.

Combination immunotherapy and novel forms of targeted therapy are critical to ongoing and current GBM research. Because the glioma microenvironment is so immunosuppressive, therapies that address tumor-promoting cell populations and mechanisms are promising and could be synergistic with existing agents. Chemokine and chemokine receptors important to GBM development and malignancy include CXCL12 and its receptors CXCR4 and CXCR7, and targeting those axes has shown promise in improving survival rates in glioma models and deserves ongoing study in patients.

Abbreviations

ATM	ataxia-telangiectasia-mutated
BMDC	bone marrow-derived cell
CXCL	chemokine (CXC motif) ligand
CXCR	chemokine (CXC motif) receptor
GAM	glioma-associated macrophage
GBM	glioblastoma
GSC	glioma stem cell
HGG	high-grade glioma
MDSC	myeloid-derived suppressor cell
MGMT	methylguanine-DNA-methyltransferase

Immunotherapeutic Strategies for the Treatment of Glioma
https://doi.org/10.1016/B978-0-12-819755-4.00009-6

SGZ subgranular zone
siRNA small interfering RNA
SVZ subventricular zone
VEGF vascular endothelial growth factor

Conflict of interest

The authors whose names are listed certify that they have no affiliations or involvement with any organization or entity with any financial interest in the subject matter or materials discussed in this manuscript.

Introduction

Gliomas are the most common central nervous system malignancies and are associated with poor prognoses, despite aggressive surgical management. Glioblastoma (GBM) is the most aggressive and common glioma, and remains difficult to treat, with 1-year and 2-year mortality rates of approximately 50% and 75%, respectively [1]. The combination of temozolomide, a DNA-alkylating agent, with postoperative external beam radiotherapy has improved outcomes [2]. In addition, combination therapies involving temozolomide with novel types of treatment, such as the application of alternating electric fields, have also shown promise in recent trials [3]. DNA alkylation is most effective in treating tumors with methylation of the *O*-6-methylguanine-DNA-methyltransferase (MGMT) promoter, in which MGMT protein expression is suppressed and dealkylation of DNA is impaired [4]. In patients with unmethylated *MGMT*, which constitute approximately half of all HGGs, postoperative temozolomide, along with older-generation alkylating agents, has generally yielded poor outcomes [4, 5]. Additionally, managing recurrent gliomas remains challenging, as no effective treatments currently exist. Improved therapies are needed for this devastating disease.

Immunotherapy for gliomas

Recent discoveries in immunology and increased understanding of its role in the tumor microenvironment have led to the development of numerous FDA-approved immunotherapy treatments which are unified by their ability to potentiate or direct an immunological response against the tumor [6]. Multiple subclasses of immunotherapeutic agents have been studied in the context of gliomas, the foremost of which are monoclonal antibodies aimed at two of the hallmark characteristics of tumor growth: immune suppression and neovascularization.

However, no survival benefits have been identified in large-scale trials of immunotherapy in GBM. A phase III trial (Checkmate-498) comparing nivolumab, a PD-1 monoclonal antibody, to temozolomide in newly diagnosed, *MGMT*-unmethylated GBM did not reveal a survival benefit associated with receipt of immunotherapy (NCT02617589; publication pending). The EORTC 26101 trial, which evaluated the benefit of *VEGF* inhibition in progressive GBM receiving lomustine, also did not uncover a survival benefit; furthermore, a higher rate of adverse events was observed in patients receiving additional immunotherapy [7]. These results, particularly in the context of successful trials in other cancers, suggest that GBM poses a uniquely difficult challenge to immunotherapeutic approaches and hints at the importance of the glioma tumor microenvironment in modulating immune responses to therapy.

Glioma and GBM microenvironment

Despite the sobering results of immunotherapy for HGGs so far, indications supporting the continued exploration of immunotherapies in HGGs are numerous. Foremost is the characterization of the HGG tumor microenvironment, including the discovery of cellular niches established by glioma stem cells (GSCs), suggesting a prominent role of immune subpopulations in maintaining tumorigenicity. In GBM, GSCs act through multiple mechanisms. Periostin-regulated recruitment of tumor-supportive M2 macrophages by GSCs has been associated with dysfunction of the local antitumor immune response and poor clinical prognosis [8–12]. GSCs also exhibit proangiogenic tendencies through an increased production of *VEGF* and subsequent establishment of tumor-derived endothelial-like cells promoting vasculogenesis [13–16]. In response to DNA damage induced by ionizing radiation, GSCs preferentially induce DNA damage checkpoint responses through the increased activation of ataxia-telangiectasia-mutated (ATM), Chk1/2, and Rad17, potentiating downstream DNA repair and conferring radioresistance to the tumor [17].

Chemokine receptors, chemokines, and gliomas

Emerging evidence suggests that GSC activity is modulated by chemokine signaling, contributing to GSC-mediated proangiogenic effects and GSC invasion of the subventricular zone (SVZ) and subgranular zone (SGZ) [18, 19]. These regions may contain the glioma cells most resistant to modern chemotherapy and radiation, suggesting that disruption of either the formation or propagation of the SVZ and SGZ niches would be promising for future therapies.

Chemokines interact with their receptors to facilitate cellular movements of not only hematopoietic and immune cells but also malignant tumor cells, which also possess certain chemokine receptors. This important interplay within the multitude of chemokine axes has become a focus within glioma and GBM research. Chemokines contribute to neural development, cellular proliferation, migration, and angiogenesis—different processes that gliomas can harness. As a result, chemokine and their receptors are attractive and novel targets in glioma immunotherapy.

Combination immunotherapy for gliomas

While most efforts to assess the potential of immunotherapies have relied on monotherapies (or combination therapies consisting of a single immunotherapeutic agent), recent forays into combination immunotherapy suggest tantalizing possibilities. In particular, the dual blockade of chemokine receptors and the PD-1/PD-L1 axis have yielded promising preclinical results. Inhibition of PD-1/PD-L1 interaction contributes to potentiating the CD8+ T-cell antitumoral response; however, immunosuppressive mechanisms have been identified that may temper the response to single-agent checkpoint blockade. The role of chemokine receptors and their ligands in the recruitment and activation of immunosuppressive modulators is well documented [21–23]. Additionally, chemokine receptor-ligand axes have been associated with increased tumor neoangiogenesis [23].

Preclinical studies have suggested that chemokine receptor antagonists could provide synergistic value to PD-1/PD-L1 blockade in GBM [24]. Similarly, investigational studies have posited that the addition of chemokine receptor antagonists could amplify the effect of *VEGR* inhibitors [25]. Additional validation of safety in human patients is necessary; however, as efforts to characterize the response of HGGs to immunotherapy accelerate, multiaxis targeting represents a logical next step in clinical and scientific exploration.

This book chapter aims to characterize the various chemokine receptors and chemokines, with a focus on those with therapeutic significance in recent or current studies, important to glioma and GBM growth and progression. Several chemokine receptors are also targets of therapeutic antagonists, with some showing promise in ongoing clinical studies.

CXCR4/CXCL12

Perhaps the most studied and well-known chemokine receptors involved in glioma and GBM is CXCR4 that interacts with its ligand CXCL12/SDF-1 in a variety of critical cellular and immunologic processes. CXCR4 is not only expressed in T and B lymphocytes, myeloid cells, macrophages, and endothelial cells, but also in over 23 types of cancers, including GBM [26, 27]. CXCR4 overexpression correlates with tumor proliferation and progression, metastasis, tumor-promoting angiogenesis, and treatment resistance [26].

The glioma microenvironment is characterized by populations of immunosuppressive cells, and immunotherapies may be instrumental in modulating this protumor milieu for treatment efficacy. One such population is glioma-associated macrophages and microglia (GAMs), which include the antitumor M1 polarization and the protumor M2 polarization. CXCR4 is known to be expressed on endothelial cells in tumors with a high level of immunosuppressive M2 GAMs [28]. Targeting CXCR4 decreased the numbers of CD11b+ GAMs and subsequently shifted the polarization of GAMs to the tumor-suppressing M1 form in a murine GBM model [29]. Another group of immunosuppressive cells, myeloid-derived suppressor cells (MDSCs), also feature prominently within the glioma microenvironment [30]. In both glioma and ovarian cancer murine models, the combination treatment of anti-PD-1 with a small-molecule anti-CXCR4 agent decreased the levels of MDSCs and contributed to improved overall survival rates [24, 31].

CXCR4 has a critical role in angiogenesis, a process harnessed by tumors to support aberrant growth and spread. Glioma stem cells (GSCs), which express high levels of CXCR4 in both mRNA and protein, localize to regions proximal to glioma vasculature [19]. By inhibiting CXCR4/CXCL12 axis-mediated regulation of immunosuppressive cell and tumor migration as well as angiogenesis through the Akt signaling pathway, the invasiveness of glioma tumors may be curtailed [19, 32]. As a result, tumor-promoting VEGF production was also reduced [19]. Synergistic effects between anti-VEGF immunotherapies, such as bevacizumab, and CXCR4 blockade have been noted in preclinical glioblastoma studies, resulting in a reduction of intracranial tumor burden and prolonged survival [25, 33]. A corollary effect of targeting CXCR4 may involve reducing GSC movement itself, leading to an overall improvement in survival in murine models. In addition, the distinct process of vasculogenesis, which refers to the creation of vasculature by hematopoietic cells from the bone marrow, is blocked by targeting CXCR4 [27]. In a study illustrating a CXCR4 antibody's effect on vasculogenesis, inhibiting the CXCR4 axis led to downregulation of bone marrow-derived cell (BMDC) migration into recurrent gliomas and subsequent prevention of aberrant tumor vasculature regrowth [27].

Combination immunotherapy involving CXCR4 antagonists has demonstrated promise in preclinical studies. By addressing different aspects of the immunological mechanisms of tumor suppression, anti-CXCR4 agents may potentiate the preexisting armament of therapies for glioma and GBM. In addition to synergizing with other immunotherapeutic agents, such as anti-PD-1 [24] and antiangiogenic agents [33], CXCR4 antagonists may further enhance traditional cytotoxic chemotherapy, such as BCNU, by promoting apoptosis and inhibiting tumor proliferation [34].

CXCR7/CXCL12

Also, a receptor for CXCL12, CXCR7, otherwise known as ACKR3, plays an important role in glioma progression and could potentially act as a novel treatment target as well. While the expression of chemokine receptor levels on the tumor surface is variable in compositionally heterogeneous cancers like glioma and GBM, CXCR7 expression levels, both intracellularly and on the cell surface, correlate with malignancy, prognosis, treatment failure, and poor overall survival of GBM [35–37]. Glioma samples demonstrate impressive levels of CXCR7 mRNA, demonstrating up to a ninefold enrichment in tumor tissue compared with normal brain [38]. Tumor-associated macrophages, microglia, and endothelial cells also exhibit high expression of CXCR7 [39, 40].

Given its prevalence on gliomas and GBM, CXCR7 is an attractive target for novel immunotherapeutic agents. A small-molecule antagonist of CXCR7, CCX771, reduced the transendothelial migration of CXCR7-bearing tumor cells [41]. Another group used small interfering RNA (siRNA) technology to manipulate CXCR7 protein levels in patients with glioma [38]. Introduction of the siRNA to knock down CXCR7 production directly led to the inhibition of ERK1/2 phosphorylation by the ligand CXCL12 and subsequently downregulated glioma cell migration and proliferation.

A drive to combine different agents is transforming the landscape of glioma treatment. The effects of temozolomide, a mainstay of GBM management, may be potentiated by CXCR7 blockade [42]. By incorporating this synergistic treatment in a mouse glioma model, higher levels of GBM cell death were induced, resulting in prolonged survival in murine GBM models.

CCR5/CCL5

CCR5 and its ligand CCL5 are implicated in glioma invasion and are frequently overexpressed in GBM [43]. Found on T lymphocytes and macrophages, CCR5 overexpression may correlate with increased tumor malignancy and poor prognosis [44]. Interestingly, the same downstream pathway involved in CXCR7 signaling, ERK1/2 phosphorylation, which is inhibited with CXCR7 knockdown, is part of CCR5/CCL5 signaling as well [45, 46]. CCR5/CCL5 also leads to the activation of the Akt pathway and subsequently promotes GBM progression and treatment resistance [44].

Based on the mechanisms of the CCR5 axis in tumor promotion, several studies have sought to characterize the results and efficacy of CCR5 blockade. So far, a CCR5/CCR1 antagonist decreased tumor growth in a breast cancer model [47]. One group who used a murine glioma model showed that the reduction of CCR5 by knockdown subsequently led to a decrease in tumor burden [44]. Maraviroc, a known CCR5 antagonist used in the treatment of HIV infection, prevented the development of the tumor-promoting M2 phenotype in rat microglia cultured in glioma cell media [48].

CCR2/CCL2

CCL2 is a chemokine produced by macrophages and microglia, and is heavily involved in the mobilization of GAMs to proinflammatory and tumor-bearing regions [49]. CCL2 levels are correlated with higher tumor malignancy and poor prognosis [50, 51]. As CCR2 binds with CCL2, immunosuppressive populations of MDSCs are recruited to the glioma microenvironment [22]. CCL2 also interacts with CCR4, and blockade of CCR4 reduced immunosuppressive regulatory T cells and improved survival in a glioma model [22]. By knocking down CCL2 with siRNA, angiogenesis and tumor proliferation in a glioma cell line were downregulated [52]. Another group synthesized a fusion protein aimed to block CCR2 and the subsequent downstream phosphorylation of STAT3, thereby reducing tumor burden in a murine medulloblastoma model [53]. No glioma models are currently undergoing study with regard to the CCR2/CCL2 chemokine axis.

CXCR2/IL-8

CXCR2 interacts with interleukin 8 (IL-8) and primarily activates neutrophils and perpetuates inflammation, angiogenesis, and tumorigenesis [54]. CXCR2 levels correlated with glioma grade and tumor recurrence rates [54]. Likewise, IL-8 was also correlated with recurrent GBM and inversely associated with GBM patient survival [55, 56]. Through NF-κB-Snail signaling pathway, IL-8 may influence glioma cell invasiveness [56, 57]. Utilizing a CXCR2 inhibitor, SB225002, resulted in decreased tumor proliferation and neovascularization, which could combat glioma treatment resistance to current antiangiogenic agents [58].

Future potential of chemokine receptor antagonists for glioma treatment

While a range of preclinical studies have examined the efficacy of chemokine receptor antagonist immunotherapy for glioma and GBM, these agents deserve further investigation in patient clinical trials. Some preclinical studies have utilized siRNA technology or other methods to knock down certain chemokine receptor expression. However, the agents showing most promise for translating into patient studies are small-molecule antibodies against chemokine receptors involved in glioma malignancy and invasiveness.

So far, several chemokine receptor antagonists are undergoing clinical trial investigation for various cancers. So far, the majority of trials for GBM involve various anti-CXCR4 forms as either stand-alone therapy or in combination with bevacizumab and sunitinib or anti-VEGF therapy [32, 59, 60]. Unfortunately, another clinical trial investigating the safety and efficacy of plerixafor, an antagonist against CXCR4 as well, in combination with avastin was terminated due to low patient enrollment (NCT01339039). Recht et al. studied various doses of plerixafor administered to HGG patients who have already started receiving the standard temozolomide chemotherapy and radiation (NCT01977677). While longitudinal results of the studies are still pending, there were only a few incidences of adverse effects, such as encephalopathy and disease progression, particularly at the high dose of plerixafor. A separate study with the addition of whole-brain radiation to standard chemoradiotherapy and plerixafor is currently underway as well (NCT03746080).

Conclusion

Chemokine receptor antagonists are an exciting area of glioma and GBM treatment research currently. The invasiveness and treatment resistance of these deadly tumors stump many clinicians and investigators. With classic checkpoint inhibitors showing equivocal results in human studies, such as CheckMate-143 (NCT0201717), attention has been turned to alternative avenues, either as robust stand-alone therapies or as synergistic supplements to existing immunotherapeutic agents for glioma and GBM.

The most studied chemokine and receptor axes may be CXCL12 and its receptors CXCR4 and CXCR7. These two chemokine receptors are significantly expressed on tumor cells as well as endothelial cells and immunosuppressive cell populations. Studies on glioma models have shown promising results and improvements in overall survival. While there are a few human clinical trials on chemokine receptor antagonists with pending results, it would certainly be worthwhile to investigate these agents further.

References

[1] Brown TJ, et al. Association of the extent of resection with survival in glioblastoma: a systematic review and meta-analysis. JAMA Oncol 2016;2(11):1460–9.
[2] Stupp R, et al. Radiotherapy plus concomitant and adjuvant temozolomide for glioblastoma. N Engl J Med 2005;352(10):987–96.
[3] Stupp R, et al. Effect of tumor-treating fields plus maintenance temozolomide vs maintenance temozolomide alone on survival in patients with glioblastoma: a randomized clinical trial. JAMA 2017;318(23):2306–16.
[4] Hegi ME, et al. MGMT gene silencing and benefit from temozolomide in glioblastoma. N Engl J Med 2005;352 (10):997–1003.
[5] Esteller M, et al. Inactivation of the DNA-repair gene MGMT and the clinical response of gliomas to alkylating agents. N Engl J Med 2000;343(19):1350–4.
[6] Lim M, et al. Current state of immunotherapy for glioblastoma. Nat Rev Clin Oncol 2018;15(7):422–42.
[7] Wick W, et al. Lomustine and bevacizumab in progressive glioblastoma. N Engl J Med 2017;377(20):1954–63.
[8] Komohara Y, et al. Possible involvement of the M2 anti-inflammatory macrophage phenotype in growth of human gliomas. J Pathol 2008;216(1):15–24.

[9] Morford LA, et al. Apoptotic elimination of peripheral T lymphocytes in patients with primary intracranial tumors. J Neurosurg 1999;91(6):935–46.
[10] Staudt ND, et al. Myeloid cell receptor LRP1/CD91 regulates monocyte recruitment and angiogenesis in tumors. Cancer Res 2013;73(13):3902–12.
[11] Wu A, et al. Glioma cancer stem cells induce immunosuppressive macrophages/microglia. Neuro-Oncology 2010;12(11):1113–25.
[12] Zhou W, et al. Periostin secreted by glioblastoma stem cells recruits M2 tumour-associated macrophages and promotes malignant growth. Nat Cell Biol 2015;17(2):170–82.
[13] Bao S, et al. Stem cell-like glioma cells promote tumor angiogenesis through vascular endothelial growth factor. Cancer Res 2006;66(16):7843–8.
[14] Cheng L, et al. Glioblastoma stem cells generate vascular pericytes to support vessel function and tumor growth. Cell 2013;153(1):139–52.
[15] Ricci-Vitiani L, et al. Tumour vascularization via endothelial differentiation of glioblastoma stem-like cells. Nature 2010;468(7325):824–8.
[16] Wang R, et al. Glioblastoma stem-like cells give rise to tumour endothelium. Nature 2010;468(7325):829–33.
[17] Bao S, et al. Glioma stem cells promote radioresistance by preferential activation of the DNA damage response. Nature 2006;444(7120):756–60.
[18] Goffart N, et al. Adult mouse subventricular zones stimulate glioblastoma stem cells specific invasion through CXCL12/CXCR4 signaling. Neuro-Oncology 2015;17(1):81–94.
[19] Ping YF, et al. The chemokine CXCL12 and its receptor CXCR4 promote glioma stem cell-mediated VEGF production and tumour angiogenesis via PI3K/AKT signalling. J Pathol 2011;224(3):344–54.
[20] Chen J, et al. A restricted cell population propagates glioblastoma growth after chemotherapy. Nature 2012;488 (7412):522–6.
[21] Arlauckas SP, et al. In vivo imaging reveals a tumor-associated macrophage-mediated resistance pathway in anti-PD-1 therapy. Sci Transl Med 2017;9(389).
[22] Chang AL, et al. CCL2 produced by the glioma microenvironment is essential for the recruitment of regulatory T cells and myeloid-derived suppressor cells. Cancer Res 2016;76(19):5671–82.
[23] Zou W. Immunosuppressive networks in the tumour environment and their therapeutic relevance. Nat Rev Cancer 2005;5(4):263–74.
[24] Wu A, et al. Combination anti-CXCR4 and anti-PD-1 immunotherapy provides survival benefit in glioblastoma through immune cell modulation of tumor microenvironment. J Neuro-Oncol 2019;143(2):241–9.
[25] Gravina GL, et al. The brain-penetrating CXCR4 antagonist, PRX177561, increases the antitumor effects of bevacizumab and sunitinib in preclinical models of human glioblastoma. J Hematol Oncol 2017;10(1):5.
[26] Chatterjee S, Behnam Azad B, Nimmagadda S. The intricate role of CXCR4 in cancer. Adv Cancer Res 2014;124:31–82.
[27] Kioi M, et al. Inhibition of vasculogenesis, but not angiogenesis, prevents the recurrence of glioblastoma after irradiation in mice. J Clin Invest 2010;120(3):694–705.
[28] Choi SH, et al. Tumour-vasculature development via endothelial-to-mesenchymal transition after radiotherapy controls CD44v6(+) cancer cell and macrophage polarization. Nat Commun 2018;9(1):5108.
[29] Mercurio L, et al. Targeting CXCR4 by a selective peptide antagonist modulates tumor microenvironment and microglia reactivity in a human glioblastoma model. J Exp Clin Cancer Res 2016;35:55.
[30] Gielen PR, et al. Elevated levels of polymorphonuclear myeloid-derived suppressor cells in patients with glioblastoma highly express S100A8/9 and arginase and suppress T cell function. Neuro-Oncology 2016;18 (9):1253–64.
[31] Zeng Y, et al. Dual blockade of CXCL12-CXCR4 and PD-1-PD-L1 pathways prolongs survival of ovarian tumor-bearing mice by prevention of immunosuppression in the tumor microenvironment. FASEB J 2019;33(5):6596–608.
[32] Gagner JP, et al. Multifaceted C-X-C chemokine receptor 4 (CXCR4) inhibition interferes with anti-vascular endothelial growth factor therapy-induced glioma dissemination. Am J Pathol 2017;187(9):2080–94.
[33] Barone A, et al. Combined VEGF and CXCR4 antagonism targets the GBM stem cell population and synergistically improves survival in an intracranial mouse model of glioblastoma. Oncotarget 2014;5(20):9811–22.
[34] Redjal N, et al. CXCR4 inhibition synergizes with cytotoxic chemotherapy in gliomas. Clin Cancer Res 2006;12 (22):6765–71.

[35] Liu C, et al. Expression and functional heterogeneity of chemokine receptors CXCR4 and CXCR7 in primary patient-derived glioblastoma cells. PLoS ONE 2013;8(3), e59750.

[36] Deng L, et al. Chemokine receptor CXCR7 is an independent prognostic biomarker in glioblastoma. Cancer Biomark 2017;20(1):1–6.

[37] Fan H, et al. Prognostic significance of CXCR7 in cancer patients: a meta-analysis. Cancer Cell Int 2018;18:212.

[38] Liu Y, Carson-Walter E, Walter KA. Targeting chemokine receptor CXCR7 inhibits glioma cell proliferation and mobility. Anticancer Res 2015;35(1):53–64.

[39] Hattermann K, et al. Chemokine expression profile of freshly isolated human glioblastoma-associated macrophages/microglia. Oncol Rep 2014;32(1):270–6.

[40] Hattermann K, et al. The chemokine receptor CXCR7 is highly expressed in human glioma cells and mediates antiapoptotic effects. Cancer Res 2010;70(8):3299–308.

[41] Zabel BA, et al. Elucidation of CXCR7-mediated signaling events and inhibition of CXCR4-mediated tumor cell transendothelial migration by CXCR7 ligands. J Immunol 2009;183(5):3204–11.

[42] Salazar N, et al. A chimeric antibody against ACKR3/CXCR7 in combination with TMZ activates immune responses and extends survival in mouse GBM models. Mol Ther 2018;26(5):1354–65.

[43] Kouno J, et al. Up-regulation of CC chemokine, CCL3L1, and receptors, CCR3, CCR5 in human glioblastoma that promotes cell growth. J Neuro-Oncol 2004;70(3):301–7.

[44] Zhao L, et al. Critical roles of chemokine receptor CCR5 in regulating glioblastoma proliferation and invasion. Acta Biochim Biophys Sin Shanghai 2015;47(11):890–8.

[45] Leach K, Charlton SJ, Strange PG. Analysis of second messenger pathways stimulated by different chemokines acting at the chemokine receptor CCR5. Biochem Pharmacol 2007;74(6):881–90.

[46] Aramori I, et al. Molecular mechanism of desensitization of the chemokine receptor CCR-5: receptor signaling and internalization are dissociable from its role as an HIV-1 co-receptor. EMBO J 1997;16(15):4606–16.

[47] Robinson SC, et al. A chemokine receptor antagonist inhibits experimental breast tumor growth. Cancer Res 2003;63(23):8360–5.

[48] Laudati E, et al. Blockade of CCR5 receptor prevents M2 microglia phenotype in a microglia-glioma paradigm. Neurochem Int 2017;108:100–8.

[49] Hinojosa AE, et al. CCL2/MCP-1 modulation of microglial activation and proliferation. J Neuroinflammation 2011;8:77.

[50] Kuratsu J, et al. Quantitative study of monocyte chemoattractant protein-1 (MCP-1) in cerebrospinal fluid and cyst fluid from patients with malignant glioma. J Natl Cancer Inst 1993;85(22):1836–9.

[51] Platten M, et al. Monocyte chemoattractant protein-1 increases microglial infiltration and aggressiveness of gliomas. Ann Neurol 2003;54(3):388–92.

[52] Lu B, et al. Effect of CCL2 siRNA on proliferation and apoptosis in the U251 human glioma cell line. Mol Med Rep 2017;16(3):3387–94.

[53] Rafei M, et al. A MCP1 fusokine with CCR2-specific tumoricidal activity. Mol Cancer 2011;10:121.

[54] Yang L, et al. Correlation of C-X-C chemokine receptor 2 upregulation with poor prognosis and recurrence in human glioma. Onco Targets Ther 2015;8:3203–9.

[55] Hasan T, et al. Interleukin-8/CXCR2 signaling regulates therapy-induced plasticity and enhances tumorigenicity in glioblastoma. Cell Death Dis 2019;10(4):292.

[56] Zhang B, et al. Autocrine IL-8 promotes F-actin polymerization and mediate mesenchymal transition via ELMO1-NF-kappaB-Snail signaling in glioma. Cancer Biol Ther 2015;16(6):898–911.

[57] Raychaudhuri B, Vogelbaum MA. IL-8 is a mediator of NF-kappaB induced invasion by gliomas. J Neuro-Oncol 2011;101(2):227–35.

[58] Angara K, et al. CXCR2-expressing tumor cells drive vascular mimicry in antiangiogenic therapy-resistant glioblastoma. Neoplasia 2018;20(10):1070–82.

[59] Yang L, et al. Blocking CXCR4-mediated cyclic AMP suppression inhibits brain tumor growth in vivo. Cancer Res 2007;67(2):651–8.

[60] Amunarriz M, et al. Seroprevalence of Chagas disease in Aguarico canton in the Ecuadorian Amazon. Rev Panam Salud Publica 2010;28(1):25–9.

Index

Note: Page numbers followed by *f* indicate figures and *t* indicate tables.

W

X

FSC
www.fsc.org
MIX
Papier | Fördert
gute Waldnutzung
FSC® C083411

Kolibri 360 GmbH
Ferdinand-Jühlke-Straße 7
99095 Erfurt, Deutschland
produktsicherheit@kolibri360.de